McGraw-Hill Educ

Because of demand, the circulation
Of this book is limited to 14 days.

GREAT NECK LIBRARY
10696692
MAY 17 2018

500 ACT Science Questions

to Know by Test Day

X DISCARDED OR
WITHDRAWN
GREAT NECK LIBRARY

Also in the McGraw-Hill Education 500 Questions Series

McGraw-Hill Education 500 ACT English and Reading Questions to Know by Test Day
McGraw-Hill Education 500 ACT Math Questions to Know by Test Day
McGraw-Hill Education 500 American Government Questions: Ace Your College Exams
McGraw-Hill Education 500 College Algebra and Trigonometry Questions: Ace Your College Exams
McGraw-Hill Education 500 College Biology Questions: Ace Your College Exams
McGraw-Hill Education 500 College Calculus Questions: Ace Your College Exams
McGraw-Hill Education 500 College Chemistry Questions: Ace Your College Exams
McGraw-Hill Education 500 College Physics Questions: Ace Your College Exams
McGraw-Hill Education 500 Differential Equations Questions: Ace Your College Exams
McGraw-Hill Education 500 European History Questions: Ace Your College Exams
McGraw-Hill Education 500 French Questions: Ace Your College Exams
McGraw-Hill Education 500 Linear Algebra Questions: Ace Your College Exams
McGraw-Hill Education 500 Macroeconomics Questions: Ace Your College Exams
McGraw-Hill Education 500 Microeconomics Questions: Ace Your College Exams
McGraw-Hill Education 500 Organic Chemistry Questions: Ace Your College Exams
McGraw-Hill Education 500 Philosophy Questions: Ace Your College Exams
McGraw-Hill Education 500 Physical Chemistry Questions: Ace Your College Exams
McGraw-Hill Education 500 Precalculus Questions: Ace Your College Exams
McGraw-Hill Education 500 Psychology Questions: Ace Your College Exams
McGraw-Hill Education 500 SAT Critical Reading Questions to Know by Test Day
McGraw-Hill Education 500 SAT Math Questions to Know by Test Day
McGraw-Hill Education 500 Spanish Questions: Ace Your College Exams
McGraw-Hill Education 500 Statistics Questions: Ace Your College Exams
McGraw-Hill Education 500 U.S. History Questions, Volume 1: Ace Your College Exams
McGraw-Hill Education 500 U.S. History Questions, Volume 2: Ace Your College Exams
McGraw-Hill Education 500 World History Questions, Volume 1: Ace Your College Exams
McGraw-Hill Education 500 World History Questions, Volume 2: Ace Your College Exams
McGraw-Hill Education 500 MCAT Biology Questions to Know by Test Day
McGraw-Hill Education 500 MCAT General Chemistry Questions to Know by Test Day
McGraw-Hill Education 500 MCAT Organic Chemistry Questions to Know by Test Day
McGraw-Hill Education 500 MCAT Physics Questions to Know by Test Day

McGraw-Hill Education

500 ACT Science Questions

to Know by Test Day

Second Edition

Anaxos, Inc.

New York Chicago San Francisco Athens London Madrid
Mexico City Milan New Delhi Singapore Sydney Toronto

GREAT NECK LIBRARY

Copyright © 2018, 2015 by McGraw-Hill Education. All rights reserved. Printed in the United States of America. Except as permitted under the United States Copyright Act of 1976, no part of this publication may be reproduced or distributed in any form or by any means, or stored in a database or retrieval system, without the prior written permission of the publisher.

1 2 3 4 5 6 7 8 9 QFR 22 21 20 19 18

ISBN 978-1-260-10830-9
MHID 1-260-10830-9

e-ISBN 978-1-260-10831-6
e-MHID 1-260-10831-7

ACT is a registered trademark of ACT, Inc., which was not involved in the production of, and does not endorse, this product.

McGraw-Hill books are available at special quantity discounts to use as premiums and sales promotions or for use in corporate training programs. To contact a representative, please visit the Contact Us pages at www.mhprofessional.com.

CONTENTS

INTRODUCTION

Congratulations! You've taken a big step toward ACT success by purchasing *McGraw-Hill Education: 500 ACT Science Questions to Know by Test Day*. This book and the others in this series were written by expert teachers who know the ACT inside and out, and working on these questions will expose you to many kinds of questions that are likely to appear on the exam. We are here to help you take the next step and score high on your ACT exam so you can get into the college or university of your choice.

This book provides 500 ACT-style, multiple-choice questions that cover a wide variety of science topics. Each question is clearly explained in the answer key. The questions will give you valuable independent practice to supplement the material you have already covered in your science classes.

You may be the kind of student who needs extra study a few weeks before the exam for a final review, or you may be the kind of student who puts off preparing until the last minute. No matter what your preparation style, you will benefit from reviewing these 500 questions, which closely parallel the content and degree of difficulty of the science questions on the actual ACT. These questions and the explanations in the answer key are the ideal last-minute study tool for those final weeks before the test.

As you go through each chapter, you'll want to practice your pacing as you read the passages and answer the questions because time is a critical factor during the ACT test. After reading or skimming the passage, a good rule of thumb is to allow yourself 30 seconds per question. Determine what pace enables you to be most successful in answering the questions. Chapter 12 provides a complete 35-minute practice test with seven passages containing 40 questions that mirror the format of the ACT test.

If you practice with all the questions and answers in this book, we are certain you will build the skills and confidence you will need to excel on the ACT. Good luck!

—*The Editors of McGraw-Hill Education*

CHAPTER 1

Test 1

Passage 1

A corn seed, or kernel, is made up of pericarp, aleurone, and endosperm layers. Figure 1.1 shows the basic anatomy of a corn seed. The endosperm layer may be yellow or white. The aleurone layer may be purple, red, or colorless. Unless the aleurone is colorless, the color of the aleurone layer masks the color of the endosperm layer.

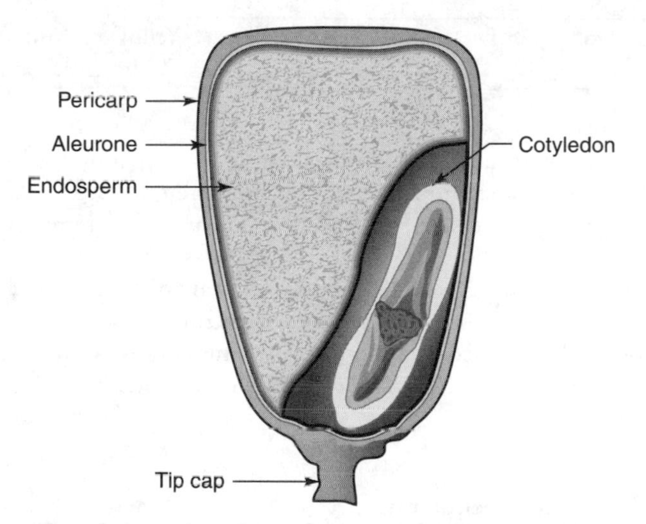

Figure 1.1

Endosperm color is determined by a single gene with two versions, or *alleles*. A corn seed's specific combination of alleles, or *genotype*, determines the physical color of the endosperm (*phenotype*). Aleurone color is determined by the interactions of three independently assorting genes. A genotype that contains at least one aleurone color allele *R* and one aleurone color inhibitor allele *C* will produce a purple aleurone. A genotype that contains at least one *R*, one *C*, and two aleurone color modifier alleles *pp* will produce a red aleurone. All other allele combinations will produce a colorless aleurone.

Table 1.1 shows the phenotypes that result from all possible allele combinations for each of the four corn seed color genes.

TABLE 1.1 Corn Seed Color

Gene	Allele Relationships	Genotype	Phenotypic Outcome
Aleurone color	$R > r$	RR or Rr	Purple aleurone in presence of C allele
		rr	Colorless aleurone
Aleurone color inhibitor	$C' > C > c$	$C'C'$, $C'C$	Colorless aleurone or $C'c$
		CC or Cc	Purple aleurone in presence of R allele
		cc	Colorless aleurone
Aleurone color modifier	$P > p$	PP or Pp	No effect on aleurone color
		pp	Changes purple aleurone to red
Endosperm color	$Y > y$	YY or Yy	Yellow endosperm
		yy	White endosperm

In a single ear of corn, each individual kernel is a separate seed representing an independent outcome from the cross of the parental corn. This means that individual kernels on the same ear of corn can have different genotypes and phenotypes.

Students in a biology class examined several ears of corn that resulted from three different parental crosses. Students were told the parental phenotypes for each cross and were instructed to count the number of kernels of each color present on each ear of corn. Table 1.2 shows the students' kernel color data for each of the three crosses.

TABLE 1.2 Corn Seed Genetic Cross

	Parental Phenotypes	Phenotypic Outcomes	
Cross 1	Yellow × white	Ear 1	503 yellow
		Ear 2	510 yellow
		Ear 3	506 yellow
Cross 2	Red × red	Ear 1	381 red; 126 yellow
		Ear 2	384 red; 124 yellow
		Ear 3	380 red; 123 yellow
Cross 3	Purple × yellow	Ear 1	256 purple; 249 yellow
		Ear 2	255 yellow; 253 purple
		Ear 3	257 yellow; 251 purple

1. Which structure's color is only visible when the aleurone layer is colorless?

(A) Pericarp
(B) Tip cap
(C) Cotyledon
(D) Endosperm

2. It can most logically be inferred that the pericarp layer of a corn seed:

(A) is colorless.
(B) is beneath the aleurone and endosperm layers.
(C) has the same phenotype as the endosperm layer.
(D) is absent in most corn seeds.

3. According to Table 1.1, which trait has more than two alleles?

(A) Aleurone color modifier
(B) Aleurone color
(C) Aleurone color inhibitor
(D) Endosperm color

4. According to Table 1.1, how many unique kernel color phenotypes are possible?

(A) Two
(B) Four
(C) Five
(D) Nine

5. Based on the information in Table 1.1, which of the following genotypes would produce a red kernel?

(A) *rrCCppyy*
(B) *rrCCPpyy*
(C) *RRCCppyy*
(D) *RRCCPpyy*

6. The term *allele relationships* describes how multiple alleles for the same gene interact. Based on the information in Table 1.1, which statement accurately describes the relationship between the alleles of the aleurone color modifier gene?

(A) When both *P* and *p* are present, an intermediate phenotype is produced.
(B) When *P* is present, the phenotype of *p* is masked.
(C) When *p* is present, the phenotype of *P* is masked.
(D) The relationship between *P* and *p* cannot be determined from the information in the table.

7. Table 1.2 shows two different kernel colors on the same ear of corn. This is possible because

(A) different kernels have different parent plants.
(B) some kernels do not have an aleurone layer.
(C) each kernel only gets two of the four seed color genes.
(D) each kernel represents a separate offspring.

8. If the genotype of the yellow parent in Cross 1 is *rrCCppYY*, which of the following could be the genotype of the white parent?

(A) *rrCCppyy*
(B) *rrccppYY*
(C) *RrCCppYY*
(D) *rrCCPPYY*

9. In Cross 2, two red parents are shown to produce yellow kernels. What is the most likely explanation for this outcome?

(A) A mutation occurring when the two parent plants were crossed resulted in a new color phenotype.
(B) The crossing of parent alleles resulted in some kernels with a colorless aleurone phenotype.
(C) The two parent plants for Cross 2 were incorrectly identified, resulting in mismatched phenotypes.
(D) One of the parent plants passed on a yellow allele instead of a red aleurone color modifier allele.

10. To have the greatest probability of producing a yellow kernel, it would be most appropriate to repeat Cross(es):

(A) 1.
(B) 3.
(C) 1 and 3.
(D) 1, 2, and 3.

11. Based on the relationships information in Tables 1.1 and 1.2, what would be the outcome if the yellow parent plant in Cross 3 were replaced with a white parent?

(A) The ratio of purple to yellow kernels would increase.
(B) The ratio of purple to yellow kernels would remain constant.
(C) The resulting ears would contain purple and white kernels.
(D) The resulting ears would contain purple, yellow, and white kernels.

12. Corn seed color is considered a polygenic trait. Based on the information in the passage, the term *polygenic* refers to a trait that

(A) results from a single gene with multiple alleles.
(B) can exhibit a variety of phenotypes over time.
(C) has a phenotype that is influenced by multiple genes.
(D) affects many different functions of an organism.

13. Most of the corn sold in grocery stores is yellow. This means that an ear of corn seen at the grocery store possesses

(A) a different combination of genes than is shown in Table 1.1.
(B) fewer color genes than the corn in the crosses shown here.
(C) the same genotype as the yellow kernels produced in Cross 2.
(D) a genotype that produces a colorless aleurone.

Passage 2

Four basic aerodynamic forces act on an airplane, whether it is a passenger jet or a model made of paper. *Thrust* is the forward force and *drag* is the backward force, both of which act parallel to the airplane's motion. *Lift* is the upward force that acts perpendicular to the airplane's motion. *Gravity* is the downward force.

Students performed three experiments to determine the effects of different physical modifications on the flying ability of paper airplanes. In each experiment, students used printer paper to create a set of identical paper dart planes. They then modified the airplanes' design to investigate the effect on flight distance.

In each experiment, a single student gently threw each airplane. A second student then measured the horizontal distance covered by each plane. The students performed each experiment three times for each airplane.

Experiment 1

Students created three identical paper airplanes. The first plane's flat wings were left unaltered. The second plane's wings were modified to curve upward in a U shape. The third plane's wings were modified to curve downward in an inverted U shape. The results are shown in Table 1.3.

TABLE 1.3 Airplane Wing Curvature Data

Wing Curvature	Horizontal Distance (m)
Trial 1	
Flat wings	9.5
Wings curved up	10.2
Wings curved down	9.1
Trial 2	
Flat wings	10.4
Wings curved up	10.3
Wings curved down	9.9
Trial 3	
Flat wings	10.8
Wings curved up	10.6
Wings curved down	10.8

Experiment 2

Students created three identical paper airplanes. They left the first plane's flat wings unaltered. The second plane's wingtips were bent slightly upward. The third plane's wingtips were bent slightly downward. The results are shown in Table 1.4.

TABLE 1.4 Airplane Wingtip Position Data

Wingtip Position	Horizontal Distance (m)
Trial 1	
Flat wingtips	10.6
Wingtips bent up	12.4
Wingtips bent down	3.2
Trial 2	
Flat wingtips	11.1
Wingtips bent up	12.9
Wingtips bent down	3.5
Trial 3	
Flat wingtips	10.9
Wingtips bent up	12.8
Wingtips bent down	3.3

Experiment 3

Students created four identical paper airplanes. The first plane remained unaltered. Two paperclips were placed on either side of the second plane's nose. Two paperclips were placed on either side of the third plane at midwing. Two paperclips were placed on either side of the fourth plane's tail. The results are shown in Table 1.5.

TABLE 1.5 Airplane Paperclip Position Data

Paperclip Position	Horizontal Distance (m)
Trial 1	
No paperclips	10.8
Nose	9.1
Midwing	10.6
Tail	6.3
Trial 2	
No paperclips	10.7
Nose	9.4
Midwing	10.5
Tail	6.2
Trial 3	
No paperclips	10.8
Nose	9.3
Midwing	10.7
Tail	6.0

14. Which aerodynamic force is the result of friction as an airplane moves through the air?

(A) Thrust
(B) Lift
(C) Drag
(D) Gravity

15. Which pair of aerodynamic forces directly oppose each other?

(A) Lift and drag
(B) Thrust and lift
(C) Gravity and thrust
(D) Drag and thrust

16. Experiments 1 and 3 differed in the:
(A) type of paper used.
(B) number of airplanes tested.
(C) number of students involved.
(D) number of trials performed.

17. Which of the following statements about Experiment 1 is most accurate?
(A) Curving the wings slightly upward appears to improve airplane performance.
(B) Altering the curvature of the wings appears to have little impact on airplane performance.
(C) Curving the wings slightly downward appears to impede airplane performance.
(D) Flat wings appear to result in poor airplane performance.

18. Which condition represents the control group in Experiment 2?
(A) Wingtips bent down
(B) Wingtips bent up
(C) Flat wingtips
(D) No wingtips

19. Based on the data from Experiment 3, which paperclip placement had the least effect on flight distance?
(A) At the tail
(B) On the wingtips
(C) At the nose
(D) Midwing

20. Based on the data from the three experiments, which of the following is the approximate average horizontal distance traveled by an unaltered airplane?
(A) 9.5 m
(B) 10.0 m
(C) 10.5 m
(D) 11.5 m

21. Which modification had the most positive effect on airplane performance?
(A) Bending wingtips slightly upward
(B) Adding paperclips to the midwing
(C) Curving wings upward
(D) Bending wingtips slightly downward

22. A student produces the graph shown in Figure 1.2. This graph best represents the data contained in:

(A) Table 1.3.
(B) Table 1.4.
(C) Table 1.5.
(D) all three tables.

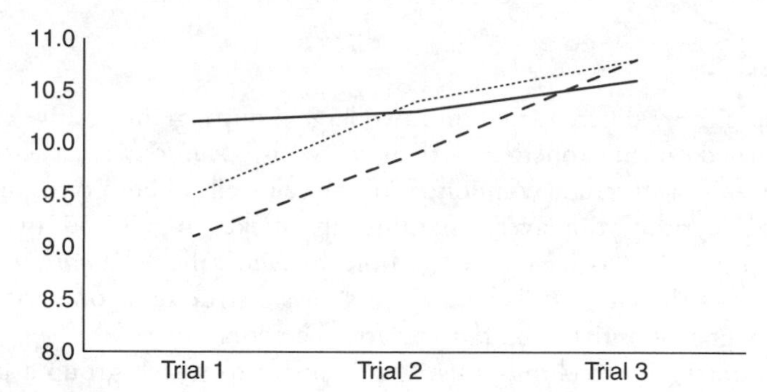

Figure 1.2

23. An object's *center of gravity* identifies the average location of the object's weight. In which experiment(s) did students alter the center of gravity of the paper airplanes?

(A) Experiments 1 and 2
(B) Experiment 2
(C) Experiment 1
(D) Experiment 3

24. Which modification changed an airplane's average horizontal distance the most?

(A) Adding paperclips to the tail
(B) Adding paperclips to the nose
(C) Curving wings downward
(D) Bending wingtips slightly downward

25. In Table 1.3, the horizontal distance of each individual airplane is shown to increase with each subsequent trial. The most reasonable explanation for this trend is that the:

(A) student measuring the distance used different meter sticks with each trial.
(B) three airplanes became more aerodynamic with each trial.
(C) student throwing increased the amount of initial thrust with each trial.
(D) effects of gravity on all three airplanes were decreased with each trial.

26. Based on the data from the three experiments, which combination of features would be expected to produce the longest flight?

(A) Wings curved down, flat wingtips, and paperclips at the nose
(B) Flat wings, wingtips bent up, and no paperclips
(C) Wings curved up, wingtips bent down, and paperclips at the tail
(D) Flat wings, flat wingtips, and no paperclips

Passage 3

Bacteria species are differentiated into two large groups, gram-positive and gram-negative, based on the properties of their cell walls. *Peptidoglycan*, a sugar–amino acid polymer, is a structural component of the cell walls of both types of bacteria, though the peptidoglycan layer is significantly thicker in gram-positive bacteria. Gram-negative bacteria have an extra lipid bilayer, called the *outer membrane*, that surrounds the entire cell. Figure 1.3 shows a structural comparison of the cell walls of gram-positive and gram-negative bacteria.

Gram staining is a technique used to identify to which group a particular bacteria species belongs based on its ability to retain a dye when rinsed with a solvent. First, the primary stain, crystal violet, is applied to the bacteria culture. An iodine solution is then added to form a complex with the crystal violet inside the cells. A decolorizer (ethyl alcohol or acetone) is added next. In gram-positive bacteria, the decolorizer dehydrates and shrinks the thick peptidoglycan layer. This traps the large crystal violet–iodine complex inside the cell, staining the cell purple. In gram-negative bacteria, the decolorizer degrades the outer membrane. This prevents the thin peptidoglycan layer from retaining the crystal violet–iodine complex, and the dye is washed out of the cell. A counterstain (safranine or fuchsin) is then added to the culture, giving decolorized gram-negative cells a red color. The counterstain is lighter colored than the primary stain, so it does not affect the outcome for gram-positive cells. After the staining procedure is completed, the treated cells are examined under a microscope to determine their color, thus identifying the group to which they belong.

Gram staining is typically the first test in a series of laboratory tests used to identify an unknown bacteria sample. Table 1.6 is a dichotomous key of characteristics that can be used to identify members of five common bacteria genera.

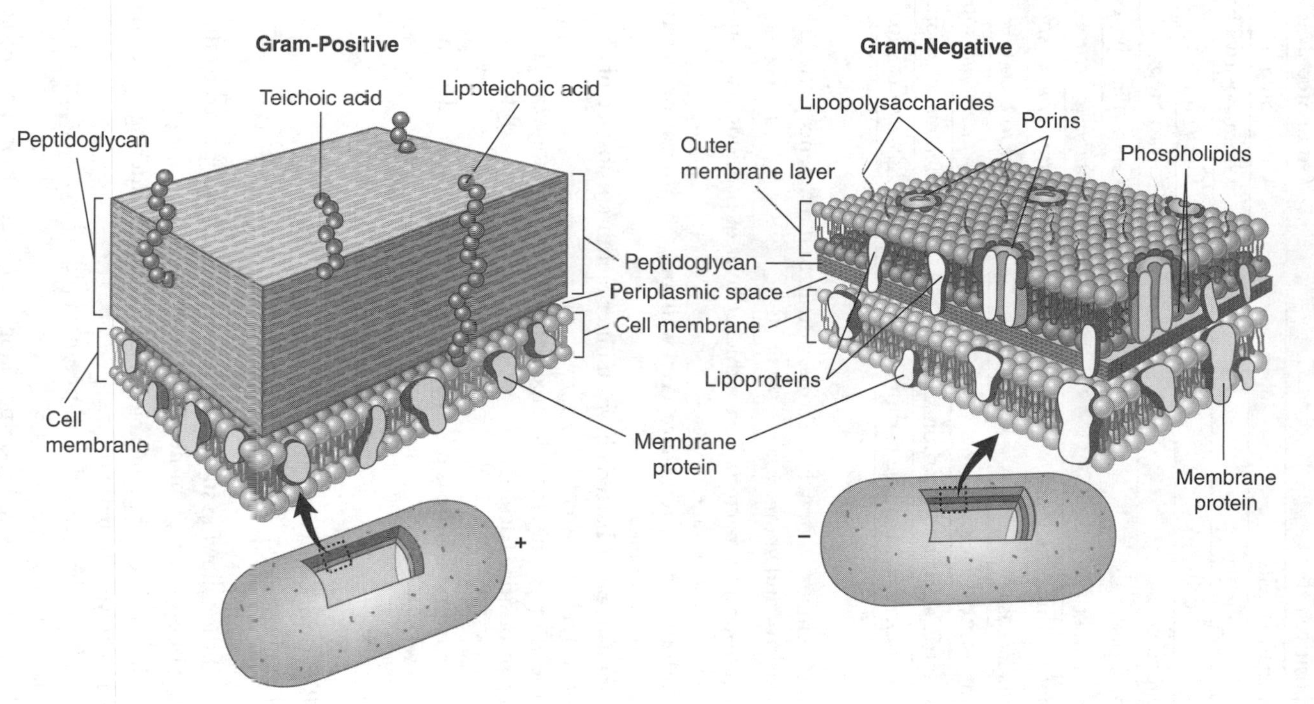
Gram-Positive
Teichoic acid
Lipoteichoic acid
Peptidoglycan
Cell
membrane
+
Gram-Negative
Lipopolysaccharides
Porins
Outer
membrane layer
Phospholipids
Peptidoglycan
Periplasmic space
Cell membrane
Lipoproteins
Membrane
protein
Membrane
protein
–

Figure 1.3

TABLE 1.6 Bacteria Dichotomous Key

1a.	Gram-positive cells	Go to Step 2
1b.	Gram-negative cells	Go to Step 3
2a.	Rod-shaped cells	Gram-positive bacilli
2b.	Sphere-shaped cells	Go to Step 4
3a.	Rod-shaped cells	Go to Step 5
3b.	Sphere-shaped cells	Gram-negative cocci
4a.	Produces catalase	*Staphylococcus* spp.
4b.	Does not produce catalase	*Streptococcus* spp.
5a.	Ferments lactose	Go to Step 6
5b.	Does not ferment lactose	*Pseudomonas* spp.
6a.	Can use citric acid as sole carbon source	*Enterobacter* spp.
6b.	Cannot use citric acid as sole carbon source	*Escherichia* spp.

27. Which statement accurately describes a structural difference between gram-positive and gram-negative bacteria?

(A) Gram-positive bacteria have a thicker layer of peptidoglycan but lack an outer membrane.
(B) Both types of bacteria have a cell wall, but gram-negative bacteria lack a cell membrane.
(C) Gram-negative bacteria have an outer membrane instead of a peptidoglycan layer.
(D) The outer membrane is located beneath the peptidoglycan layer in gram-positive bacteria.

28. Which structural feature is present in both gram-positive and gram-negative cells?

(A) Porins
(B) Lipoteichoic acid
(C) Periplasmic space
(D) Lipopolysaccharides

29. Which of the following statements is most logically supported by the presence of porins in gram-negative bacteria?

(A) Cell walls are not permeable, so all substances entering a bacteria cell must travel through porins.
(B) The lipopolysaccharide and phospholipid bilayer is less permeable than peptidoglycan.
(C) In bacteria cells, a thicker peptidoglycan layer is more permeable than a thin peptidoglycan layer.
(D) Gram-negative bacteria transport larger molecules into their cells than do gram-positive bacteria.

30. Which substance does not act as a tissue stain in the Gram staining technique?

(A) Safranine
(B) Crystal violet
(C) Fuchsin
(D) Ethyl alcohol

31. According to the Gram staining technique, a bacteria species is identified as gram-negative if its cells:

(A) appear purple after the staining procedure.
(B) have not been exposed to any stain.
(C) appear colorless after the staining procedure.
(D) appear red after the staining procedure.

32. In the Gram staining technique, which step must be performed before the addition of the iodine solution?

(A) Staining with safranine
(B) Washing with acetone
(C) Staining with crystal violet
(D) Washing with ethyl alcohol

33. Based on the information about the Gram staining technique, the most logical reason for applying a counterstain is to:

(A) intensify the appearance of gram-positive cells under a microscope.
(B) prevent the primary stain from affecting gram-negative cells.
(C) counteract the effects of the primary stain on gram-positive cells.
(D) allow gram-negative cells to be seen more easily under a microscope.

34. Based on the information about the Gram staining technique, it is most reasonable to expect a chain of which type of molecule to degrade in the presence of ethyl alcohol?

(A) Lipids
(B) Nucleotides
(C) Sugars
(D) Amino acids

35. In Table 1.6, Steps 2 and 3 list the same cell shape characteristics because:

(A) gram-positive and gram-negative bacteria can both be rod- or sphere-shaped.
(B) gram-positive bacteria can switch between rod and sphere shapes.
(C) cell shape depends on the results of the bacteria's Gram staining test.
(D) the cell shape of many gram-positive and gram-negative bacteria is unknown.

36. Of the five italicized bacteria genera listed in Table 1.6, how many have a cell wall composed of a thick peptidoglycan layer?

(A) One
(B) Two
(C) Three
(D) Five

37. Based on the information in Table 1.6, which genera contains gram-negative, rod-shaped bacteria that do not ferment lactose?

(A) *Pseudomonas*
(B) *Enterobacter*
(C) *Staphylococcus*
(D) *Escherichia*

38. Based on the information in Table 1.6, which characteristic is shared by *Pseudomonas* and *Enterobacter* bacteria?

(A) Gram-positive cells
(B) Lactose fermentation
(C) Use of citric acid as sole carbon source
(D) Rod-shaped cells

39. A laboratory technician is examining a bacteria sample belonging to the genus *Escherichia* under a microscope and notes that the sample remains colorless after performing the Gram staining procedure. It is most reasonable to assume that an error occurred during the:

(A) application of the primary stain.
(B) application of the counterstain.
(C) decolorization of the cells.
(D) bonding of iodine to the primary stain.

40. Since gram-negative bacteria are generally more resistant to antibiotics such as penicillin, Gram staining can be used to inform appropriate antibiotic treatment for patients with bacterial infections. Based on the information in Table 1.6, infections caused by bacteria belonging to which genera would be most effectively treated with penicillin?

(A) *Staphylococcus* and *Streptococcus*
(B) *Enterobacter* and *Escherichia*
(C) *Staphylococcus* and *Enterobacter*
(D) *Streptococcus* and *Escherichia*

CHAPTER 2

Test 2

Passage 4

An organism's genetic information is stored within the nuclei of its cells as a set of chromosomes. The number of chromosomes in a cell varies from species to species. In some species, the number of chromosomes can vary between individuals. Table 2.1 lists the chromosome count for a variety of species.

Ploidy is the number of sets of chromosomes present in the cell of an organism. The *monoploid* number (*x*) is the number of chromosomes an organism has in one set.

In most species, a *gamete* (sex cell) contains one complete set of an organism's chromosomes. The number of chromosomes in a gamete is referred to as the *haploid* number (*n*). The fusing of two gametes into a zygote during sexual reproduction produces *somatic cells* (body cells) containing two complete sets of chromosomes. The total number of chromosomes in a somatic cell is referred to as the *diploid* number (*2n*). In most species, the monoploid number (*x*) and the haploid number (*n*) are the same.

Some species have more than two sets of chromosomes present in their cells, a condition referred to as *polyploidy*. The somatic cells of *triploid* organisms have three sets of chromosomes, for example, and *tetraploids* have four. In polyploid organisms, the term *haploid* is still used to describe the number of chromosomes in a gamete, and *diploid* is used to describe the number of chromosomes in a somatic cell. However, the monoploid number and the haploid number are not the same in a polyploid organism.

41. Based on the information in Table 2.1, which species does not exhibit variation in chromosome numbers between individuals?

(A) European honeybee
(B) Swamp wallaby
(C) Slime mold
(D) Jack jumper ant

TABLE 2.1 Species Chromosome Count

Organism	Scientific Name	Diploid Number of Chromosomes
Adder's-tongue fern	*Ophioglossum reticulatum*	1,260
Coyote	*Canis latrans*	78
Dog	*Canis lupus familiaris*	78
Horse	*Equus ferus caballus*	64
Donkey	*Equus africanus asinus*	62
Bengal fox	*Vulpes bengalensis*	60
Silkworm	*Bombyx mori*	54
Pineapple	*Ananas comosus*	50
Zebra fish	*Danio rerio*	50
Potato	*Solanum tuberosum*	48[1]
Human	*Homo sapiens*	46
Oats	*Avena sativa*	42[2]
Mouse	*Mus musculus*	40
Earthworm	*Lumbricus terrestris*	36
Red fox	*Vulpes vulpes*	34
Alfalfa	*Medicago sativa*	32[1]
European honeybee	*Apis mellifera*	32[3]
Yeast	*Saccharomyces cerivisiae*	32
Slime mold	*Dictyostelium discoideum*	12
Swamp wallaby	*Wallabia bicolor*	10/11[4]
Fruit fly	*Drosophila melanogaster*	8
Jack jumper ant	*Myrmecia pilosula*	2[3]

[1]Organism is a tetraploid.
[2]Organism is a hexaploid.
[3]Males are haploid.
[4]Males have one less chromosome than females.
Source: http://en.wikipedia.org/wiki/List_of_organisms_by_chromosome_count.

42. The first part of an organism's scientific name identifies the *genus* to which it belongs. Which statement about the members of a genus is best supported by the information in Table 2.1?

(A) An organism's genus determines the number of chromosomes it has.
(B) Organisms in the same genus tend to have similar chromosome counts.
(C) No two organisms in the same genus can have the same number of chromosomes.
(D) Chromosome count can vary greatly between organisms in the same genus.

43. Based on the information in Table 2.1, the relationship between diploid chromosome count and organism complexity can best be described as exhibiting:

(A) a direct correlation.
(B) no correlation.
(C) an inverse correlation.
(D) a linear correlation.

44. To which kingdom does the organism exhibiting the greatest diploid number of chromosomes in Table 2.1 belong?

(A) Animalia
(B) Plantae
(C) Eubacteria
(D) Protista

45. Which species has more chromosomes than a human but fewer chromosomes than a dog?

(A) *Bombyx mori*
(B) *Canis latrans*
(C) *Ophioglossum reticulatum*
(D) *Mus musculus*

46. Cells from which pair of organisms have the same number of chromosomes in their nuclei?

(A) Horse and donkey
(B) Zebra fish and pineapple
(C) Earthworm and European honeybee
(D) Oats and potato

47. Based on the information in the passage, which species produces gametes that each contain 32 chromosomes?

(A) *Apis mellifera*
(B) *Saccharomyces cerivisiae*
(C) *Equus ferus caballus*
(D) *Drosophila melanogaster*

48. According to the information in Table 2.1, how many more total chromosomes does a female European honeybee have than a male?

(A) 1
(B) 2
(C) 16
(D) 32

49. Based on the information in the passage, the total number of chromosomes in a somatic cell is represented by which of the following terms?

(A) n
(B) x
(C) $2x$
(D) $2n$

50. Which statement about polyploidy is supported by the information in the passage?

(A) The number of chromosomes varies among the somatic cells of a polyploid organism.
(B) The gametes and somatic cells of a polyploid organism contain the same number of chromosomes.
(C) The gametes of a polyploid organism contain more than one complete set of chromosomes.
(D) The somatic cells of a polyploid organism contain too many chromosomes to be considered diploid.

51. Which of the following organisms has four complete sets of chromosomes in its somatic cells?

(A) Alfalfa
(B) Slime mold
(C) Oats
(D) Earthworm

52. Table 2.1 identifies the oat species *Avena sativa* as a hexaploid, containing six sets of chromosomes. The numerical representation $2n = 6x = 42$ describes the total number of chromosomes in a somatic cell of this hexaploid species. How many chromosomes does *Avena sativa* have in one set?

(A) 6
(B) 7
(C) 21
(D) 42

53. Which of the following correctly identifies the relationship between the diploid number ($2n$), haploid number (n), and monoploid number (x) of *Solanum tuberosum*?

(A) $2n$ is twice n, but 4 times x.
(B) $2n$ is twice the sum of n and x.
(C) $2n$ is the sum of n and x.
(D) $2n$ is twice x, but 4 times n.

Passage 5

The majority of scientists agree that global temperatures are rising, leading to a host of climate changes that will produce significant worldwide effects over time. Still subject to debate are the type and severity of effects that these climate changes will have on various industries. Two scientists present their viewpoints regarding the effects of climate change on agriculture in the United States.

Scientist 1

Climate change is likely to have mixed effects on U.S. agriculture over time. Every crop has a set of optimal conditions under which it grows and reproduces best. For many crops, the growth rate increases as temperature increases, suggesting that the progressive increase in average temperatures will have a beneficial effect on many types of crops. On the other hand, a faster growth rate means less time for the seeds of certain crops to mature, hindering their reproductive ability. Average temperatures will eventually surpass the optimal growth temperature for some crops, causing their yields to decline.

Crop yields also increase with carbon dioxide levels. The positive growth effect of carbon dioxide can be suppressed, however, if the optimal growth temperature is surpassed. The potential effects of climate change on other environmental conditions, including soil moisture, nutrient levels, and water availability must be taken into account as well.

Scientist 2

Agriculture in the United States will be adversely affected by climate change over the next several decades. Many weeds, pests, and fungi thrive in warm, wet climates and with increased levels of carbon dioxide. As average temperatures continue to increase and these conditions become more widespread, the habitat ranges for these organisms will spread northward. This will pose challenges to northern crops that have not previously been exposed to certain competitors and pests.

The predicted increase in extreme weather events will also negatively impact crop yields. An increase in the frequency of floods will destroy crops and potentially deter farming along major waterways altogether. In areas in which drought conditions are projected to become more common, a water supply capable of sustaining even modest crop yields is a very real concern.

54. According to Scientist 1, how will a change in average temperature affect the growth rates of crops?

(A) As average temperature increases, all crops will begin to grow faster.
(B) A change in average temperature will benefit some crops and harm others.
(C) If average temperature changes too quickly, many crops will stop growing.
(D) An increase in average temperature will hinder growth until crops adapt.

55. If Scientist 1 is correct, which of the following trends will most likely occur over the next several decades?

(A) The agriculture industry will experience no significant change in crop yields.
(B) The depletion of soil nutrients will cause yields of all crops to decline.
(C) Crops with chemical defenses against pests will exhibit increased yields.
(D) Crops with higher optimal growth temperatures will produce greater yields.

56. Which environmental change was discussed by Scientist 2, but not Scientist 1?

(A) Elevated carbon dioxide levels
(B) Increasing average temperatures
(C) Limited water availability
(D) Increased frequency of flooding

57. Scientist 2 did not predict that climate change would cause an increase in which of the following factors affecting crop yields?

(A) Fungi
(B) Pests
(C) Seeds
(D) Weeds

58. Based on the passage, the major difference between the opinions of Scientists 1 and 2 is that:

(A) Scientist 2 does not predict any positive effects of climate change on agriculture.
(B) Scientist 1 discusses the effects of increased temperature but not carbon dioxide.
(C) Scientist 2 expects agriculture in southern areas to be unaffected by climate change.
(D) Scientist 1 focuses only on the effects of climate change on crop reproductive rates.

59. Scientist 2 states that high:

(A) carbon dioxide levels will benefit crop yields.
(B) carbon dioxide levels will lead to decreased crop yields.
(C) average temperatures will improve crop yields.
(D) average temperatures will hinder the growth of fungi.

60. An industry-wide increase in agricultural pesticide use over the next several decades would support the opinion of:

(A) Scientist 1.
(B) Scientist 2.
(C) both scientists.
(D) neither scientist.

61. Based on the information in the passage, both scientists would agree with which of the following statements?

(A) The greatest threat posed by climate change to the U.S. agriculture industry is the projected increase in extreme weather events.
(B) Southern crops are better adapted than northern crops to withstand the effects of elevated carbon dioxide levels associated with climate change.
(C) The effects of climate change will have a greater negative impact on the reproductive ability of crops than on their growth rate.
(D) Increasing average temperatures associated with climate change will provide an advantage to some organisms.

62. It can be inferred that Scientist 1 believes elevated levels of carbon dioxide will directly lead to crops with a(n):

(A) shortened growing season.
(B) higher optimal growth temperature.
(C) decreased need for soil nutrients.
(D) increased rate of photosynthesis.

63. Which of the following does Scientist 2 identify as potential competitors to northern crops?

(A) Invasive species of weeds
(B) Newly introduced crop species
(C) Other industries that use land
(D) Migrating pest species

64. The hypothesis of Scientist 1 could best be tested by recording data over the next decade on:

(A) crop yields, average temperatures, and soil nutrient availability worldwide.
(B) seed production, soil nutrient availability, and water availability worldwide.
(C) seed production, carbon dioxide levels, and water availability in the United States.
(D) crop yields, average temperatures, and carbon dioxide levels in the United States.

65. If Scientist 2 is correct, over time, the range of:

(A) northern crops will become narrower.
(B) southern crops will move farther south.
(C) northern crops will overtake southern crops.
(D) southern crops will remain constant.

66. If Scientist 1's hypothesis is correct, which of the following graphs best represents how high carbon dioxide levels will affect crop yields over time?

(A)

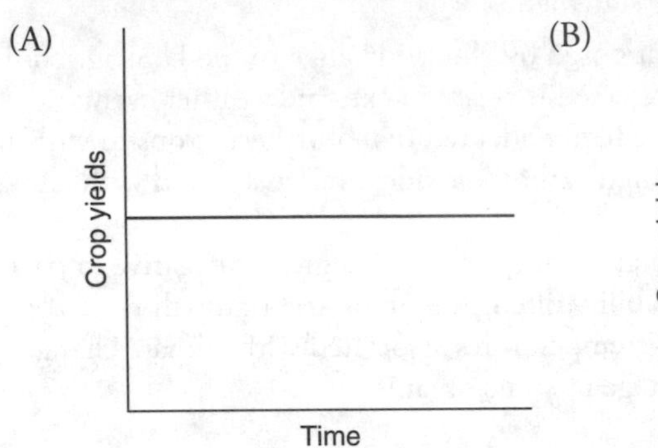

Figure 2.1

(B)

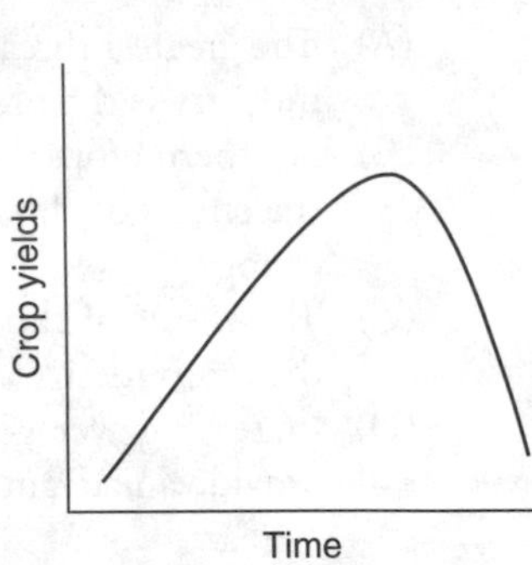

Figure 2.2

(C) (D)

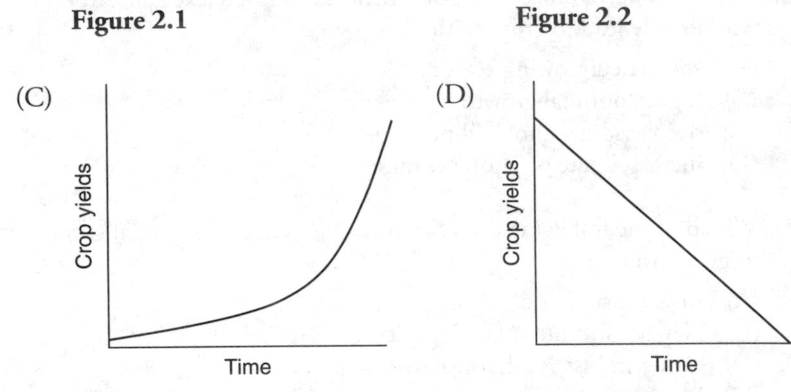

Figure 2.3

Figure 2.4

Passage 6

The leaves of green plants use the energy in sunlight to convert atmospheric carbon into organic carbon through the reactions of photosynthesis. These reactions can be summarized by the following equation:

$$6CO_2 + 6H_2O + \text{light} \rightarrow C_6H_{12}O_6 + 6O_2$$

Gas exchange between the leaf and the environment is an integral part of the photosynthesis reactions. As carbon dioxide enters the leaf, the oxygen produced as a by-product of photosynthesis is released into the environment in a 1:1 ratio. Enclosing a leaf within a lighted chamber allows for the rate of this exchange, and therefore the rate of photosynthesis, to be measured.

Students in a biology class used lighted chambers to measure the photosynthetic rate of leaves from four common plant species: sunflower, water hyacinth, rhoeo, and pothos. A leaf was placed inside the chamber, and a flow of air was introduced. Sensors within the chamber recorded data on light intensity (LED irradiance), carbon dioxide concentration, air temperature, and relative humidity.

The leaf was initially exposed to a constant light intensity of 300 $\mu E/m^2/s$ to stimulate photosynthesis. After this initial period, students incrementally increased the light intensity to investigate the relationship between light intensity and photosynthetic rate.

Figure 2.5 shows the light intensity (LED irradiance) over time for a chamber containing a water hyacinth.

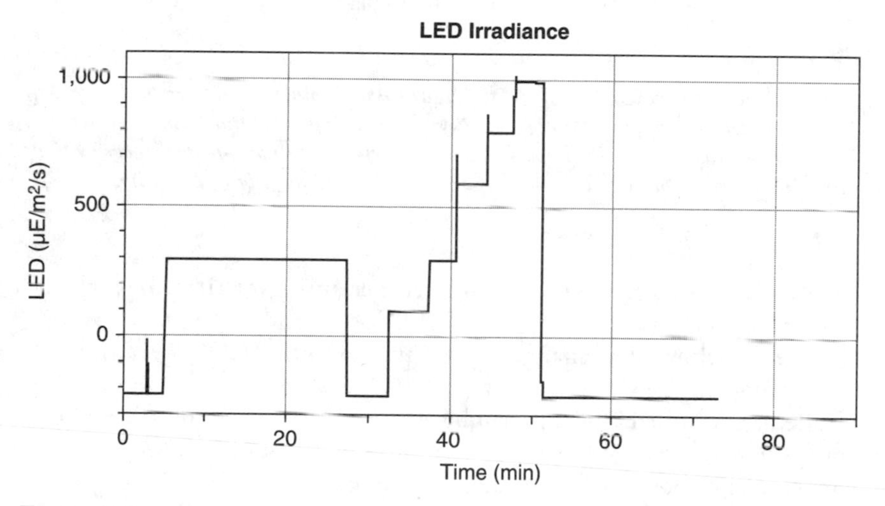

Figure 2.5

Source: "BISC 111/113: Introductory Organismal Biology," by Jocelyne Dolce, Jeff Hughes, Janet McDonough, Simone Helluy, Andrea Sequeira, and Emily A. Bucholtz. http://openwetware.org/wiki/Lab_5:_Measurement_of_Chlorophyll_Concentrations_and_Rates_of_Photosynthesis_in_Response_to_Increasing_Light_Intensity.

Figure 2.6 shows the change in carbon dioxide concentration over time for a chamber containing a water hyacinth.

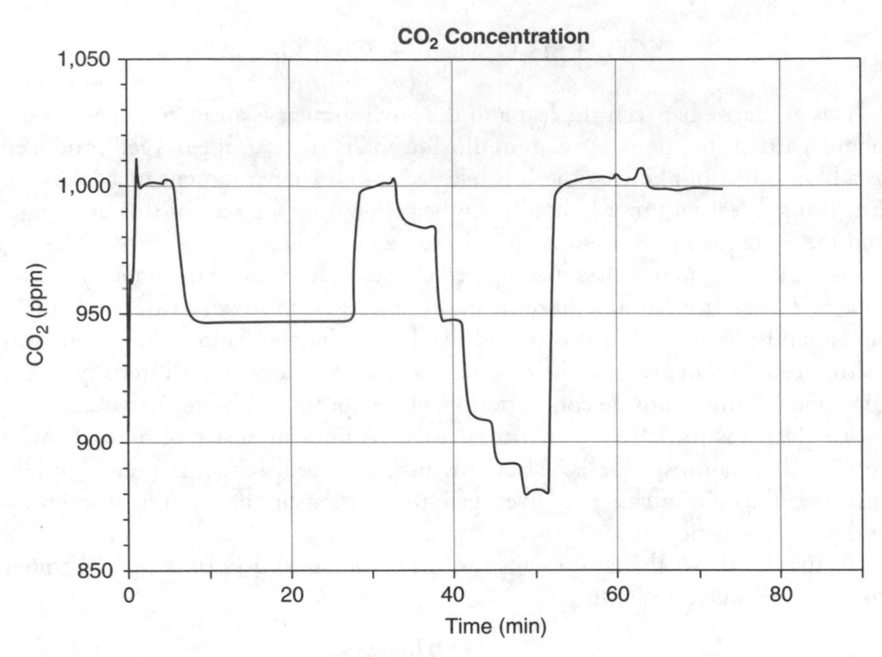

Figure 2.6

Source: "BISC 111/113: Introductory Organismal Biology," by Jocelyne Dolce, Jeff Hughes, Janet McDonough, Simone Helluy, Andrea Sequeira, and Emily A. Bucholtz. http://openwetware.org/wiki/Lab_5:_Measurement_of_Chlorophyll_Concentrations_and_Rates_of_Photosynthesis_in_Response_to_Increasing_Light_Intensity.

Figure 2.7 shows the change in air temperature over time for a chamber containing a water hyacinth.

Figure 2.8 shows the change in relative humidity (RH) over time for a chamber containing a water hyacinth.

Students performed 10 light-chamber trials with leaves from each of the four plant species. The carbon dioxide concentration data was then used to calculate the maximum carbon dioxide exchange rate for each leaf.

Table 2.2 shows the calculated and mean carbon dioxide exchange rates for each of the four plant species.

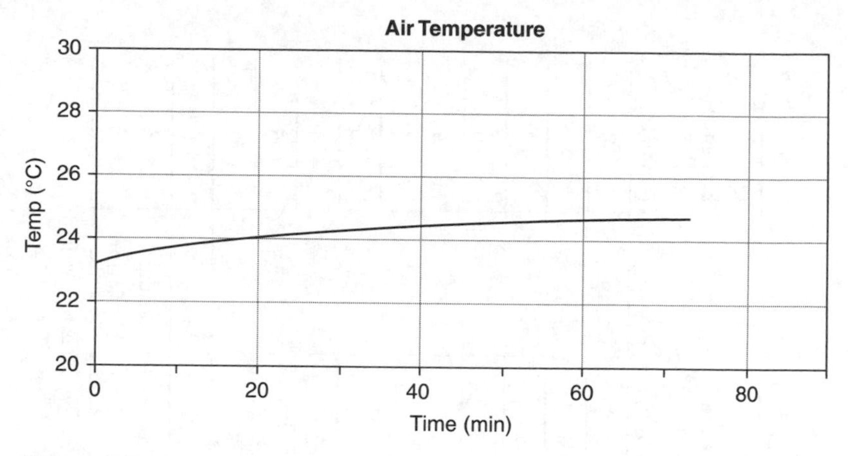

Figure 2.7

Source: "BISC 111/113: Introductory Organismal Biology," by Jocelyne Dolce, Jeff Hughes, Janet McDonough, Simone Helluy, Andrea Sequeira, and Emily A. Bucholtz. http://openwetware.org/wiki/Lab_5:_Measurement_of_Chlorophyll _Concentrations _and_Rates_of_Photosynthesis_in_Response_to_Increasing _Light_Intensity.

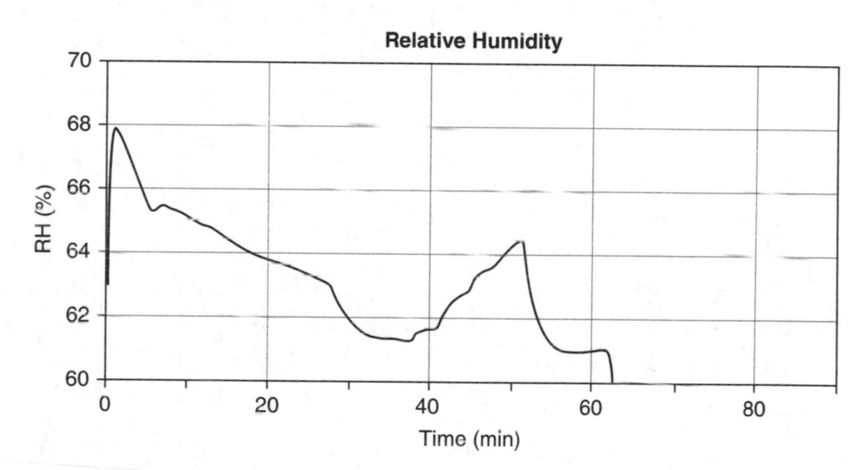

Figure 2.8

Source: "BISC 111/113: Introductory Organismal Biology," by Jocelyne Dolce, Jeff Hughes, Janet McDonough, Simone Helluy, Andrea Sequeira, and Emily A. Bucholtz. http://openwetware.org/wiki/Lab_5:_Measurement_of_Chlorophyll_Concentrations _and_Rates_of_Photosynthesis_in_Response_to_Increasing_Light_Intensity.

TABLE 2.2 Carbon Dioxide Exchange Rates

Trial	Plant	CO_2 Exchange Rate ($\mu mol/m^2/s$)	Plant	CO_2 Exchange Rate ($\mu mol/m^2/s$)	Plant	CO_2 Exchange Rate ($\mu mol/m^2/s$)	Plant	CO_2 Exchange Rate ($\mu mol/m^2/s$)
1	Sunflower	14	Water hyacinth	16	Rhoeo	2	Pothos	8
2		19		19		3		3
3		27		12		5		12
4		20		13		4		8
5		24		12		5		4
6		11		15		3		4
7		17		12		4		10
8		17		16		2		4
9		10		13		2		5
10		15		15		5		2
	Mean	17.4	Mean	14.3	Mean	3.5	Mean	6.0
	Standard Deviation	5.4	Standard Deviation	2.3	Standard Deviation	1.3	Standard Deviation	3.3

Source: http://openwetware.org/wiki/Lab_5:_Measurement_of_Chlorophyll_Concentrations_and_Rates_of_Photosynthesis_in_Response_to_ Increasing_Light_Intensity.

67. The atmospheric carbon absorbed by green plants is in the form of:
(A) carbon monoxide.
(B) carbon dioxide.
(C) carbohydrates.
(D) water.

68. Which molecule is formed as a by-product of the photosynthesis reactions?
(A) Carbon dioxide
(B) Glucose
(C) Water
(D) Oxygen

69. According to Figure 2.5, the initial photosynthesis-stimulating period lasted approximately:
(A) 5 minutes.
(B) 20 minutes.
(C) 50 minutes.
(D) 80 minutes.

70. The slight increase in air temperature indicated in Figure 2.7 is most likely related to the:
(A) increasing light intensity as the study progressed.
(B) peak in relative humidity at the 50-minute mark.
(C) increase in CO_2 concentration at the end of the study.
(D) heat generated by the sensors in the light chamber.

71. Sensors within the lighted chamber monitor the presence of which chemical reactant of the photosynthesis reactions?
(A) Carbon dioxide
(B) Oxygen
(C) Sunlight
(D) Glucose

72. Which graph represents the independent variable in the students' study?
(A) Figure 2.6
(B) Figure 2.7
(C) Figure 2.5
(D) Figure 2.8

73. Based on the data in Figures 2.5 and 2.6, which light intensity causes a water hyacinth leaf to absorb carbon dioxide at the fastest rate?

(A) 0 $\mu E/m^2/s$
(B) 300 $\mu E/m^2/s$
(C) 100 $\mu E/m^2/s$
(D) 1,000 $\mu E/m^2/s$

74. The data in Table 2.2 would best support the assertion that sunflower plants:

(A) require less intense light than the other three species.
(B) release more oxygen than the other three species.
(C) are the fastest growing of the four species studied.
(D) have the shortest life cycle of the four species studied.

75. According to Table 2.2, which plant showed the least variability across trials?

(A) Water hyacinth
(B) Pothos
(C) Rhoeo
(D) Sunflower

76. Based on the information in the passage, if the oxygen concentration within the chamber had been recorded, its graph would most closely resemble which figure?

(A) Figure 2.6
(B) Figure 2.7
(C) Figure 2.5
(D) Figure 2.8

77. According to the data in Table 2.2, which plant species perform(s) photosynthesis at a faster rate than pothos?

(A) Sunflower only
(B) Sunflower and water hyacinth
(C) Rhoeo only
(D) Rhoeo and water hyacinth

78. Which of the following generalizations is supported by the data in Figures 2.5 through 2.8?

(A) Photosynthesis occurs at a faster rate in a highly humid environment.
(B) The rate of photosynthesis varies directly with air temperature.
(C) The greater the light intensity, the faster the rate of photosynthesis.
(D) The rate of photosynthesis depends on the level of carbon dioxide available.

79. Which of the following statements is supported by the data in Table 2.2?

(A) The single leaf with the fastest gas exchange rate was from a sunflower plant.
(B) The single leaf with the slowest gas exchange rate was from a water hyacinth plant.
(C) No two leaves from different species exhibited the same gas exchange rate.
(D) No two leaves from the same species exhibited the same gas exchange rate.

80. The passage states that the rates recorded in Table 2.2 represent the maximum carbon dioxide exchange rates observed for each trial. Assuming that light intensity was increased at the same intervals for each trial, at approximately which point during each trial were the exchange rates recorded in the table most likely observed?

(A) 30 minutes
(B) 70 minutes
(C) 10 minutes
(D) 50 minutes

CHAPTER 3

Test 3

Passage 7

Antigens occur on the surface of many cell types and provide a unique chemical signature that allows the body to determine the cell's identity. *Antibodies* are proteins that attack foreign substances that may pose an immune threat to the body. Antibodies identify a substance as foreign by recognizing and binding to its surface antigens. Each type of antibody is antigen-specific, attacking only one type of antigen.

Human blood is classified into different blood groups based on the presence of certain antigens on the red blood cells. The most commonly used blood group system is ABO. This system classifies blood into four groups (types) according to the presence or absence of A and/or B antigens on the blood cells. Cells may contain A antigens only, B antigens only, both A and B antigens, or neither antigen. Blood also contains antibodies against the antigens that are absent from the red blood cells. For example, type A blood contains A antigens and anti-B antibodies. Table 3.1 identifies the antigens and antibodies present in each blood type.

TABLE 3.1 ABO Blood Types

Blood Type	Antigens Present	Antibodies Present
A	A	Anti-B
B	B	Anti-A
AB	A and B	None
O	None	Anti-A and Anti-B

Blood can also be classified as Rh-positive (Rh+) or Rh-negative (Rh–), based on the presence or absence of a different antigen on the red blood cells. Table 3.2 identifies whether the Rh antigen or antibody is present in each blood type.

TABLE 3.2 Rh Blood Types

Blood Type	Antigens Present	Antibodies Present
Rh+	Yes	No
Rh–	No	Yes

The ABO and Rh blood group systems are combined to determine an individual's medical blood type. Figure 3.1 illustrates the distribution of medical blood types in the general population of the United States.

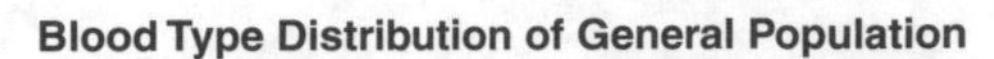

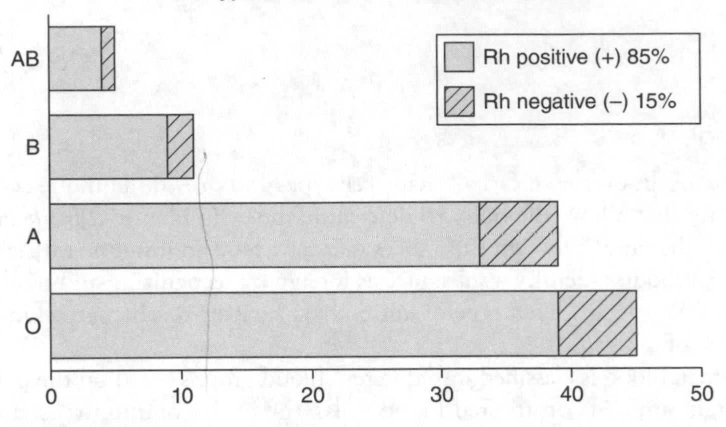

Figure 3.1

Source: https://www.armydogtags.com.

Table 3.3 indicates the distribution of medical blood types by ethnicity in the United States. The values listed represent the percentage of individuals within the given ethnic group that exhibit each blood type.

TABLE 3.3 Blood Type Demographics

	Percentage of Individuals with Blood Type (%)			
Blood Type	**Caucasian**	**African American**	**Hispanic**	**Asian**
O+	37	47	53	39
O–	8	4	4	1
A+	33	24	29	27
A–	7	2	2	0.5
B+	9	18	9	25
B–	2	1	1	0.4
AB+	3	4	2	7
AB–	1	0.3	0.2	0.1

Source: http://www.redcrossblood.org/learn-about-blood/blood-types.

81. What is the total number of medical blood types possible for a human being?

(A) Two
(B) Four
(C) Six
(D) Eight

82. The name of each ABO blood type is derived from the:

(A) antibodies that are present in the blood.
(B) antigens that are present on the red blood cells.
(C) prevalence of each blood type in the general population.
(D) antigens that are absent from the red blood cells.

83. Rh+ blood always contains:

(A) Rh antigen.
(B) anti-Rh antibodies.
(C) A and B antigens.
(D) anti-A and anti-B antibodies.

84. Blood containing anti-A and anti-Rh antibodies and B antigens would be identified as which blood type?

(A) A+
(B) B–
(C) AB–
(D) B+

85. According to Figure 3.1, what percentage of the general population has type B blood?

(A) 9%
(B) 2%
(C) 11%
(D) 16%

86. The least common blood type in the United States is type:

(A) O+.
(B) AB+.
(C) B–.
(D) AB–.

87. Based on the data in Table 3.3, which continent's population can be inferred to have the greatest incidence of blood type B+?

(A) Asia
(B) Europe
(C) Africa
(D) South America

88. In what percentage of the general U.S. population are A antigens present on red blood cells?

(A) 39%
(B) 33%
(C) 44%
(D) 37%

89. The data in Table 3.3 support the statement that more than half of the:

(A) Caucasian population has type O blood.
(B) Hispanic population has type O+ blood.
(C) general population with type O blood is Caucasian.
(D) general population with type O+ blood is Hispanic.

90. An individual of African American ethnicity has a greater chance of having a B+ blood type than:

(A) the general population.
(B) an African American has of having an A+ blood type.
(C) an individual of Asian ethnicity.
(D) an African American has of having an O+ blood type.

91. Based on the information in Table 3.1, if an individual with an AB blood type receives donated type A blood, the donated blood will cause:

(A) the conversion of existing B antigens to A antigens, altering the individual's blood type.
(B) an immune reaction because the existing B antigens will attack the new A antigens.
(C) no immune reaction because the individual has no antibodies against the new blood.
(D) the individual's body to begin producing anti-A antibodies in response to the new blood.

92. Blood type O– is often referred to as the "universal donor" because it can be donated to any of the other blood types. This is because it has:

(A) no antibodies to attack antigens.
(B) no antigens to trigger an attack by antibodies.
(C) both A and B antibodies to attack antigens.
(D) both A and B antigens to prevent attack by antibodies.

93. An individual with blood type A– can safely receive a transfusion of which of the following blood types?

(A) A+ or A–
(B) A– or AB–
(C) A– or O–
(D) O– or O+

94. The percentage of the Caucasian population that has blood type AB– is:

(A) the same as the percentage for the Hispanic population.
(B) less than the percentage for the African-American population.
(C) equal to the percentage for the general population.
(D) greater than the percentage for the general population.

Passage 8

Agarose gel electrophoresis is a technique in which an electric field is used to separate fragments of DNA by size. Figure 3.2 illustrates a common setup of an electrophoresis apparatus with the positive anode at the bottom and the cathode at the top. A square of agarose gel is prepared and placed in a tray of buffer solution. DNA in solution is loaded into small slits (*wells*) in the top of the gel. A solution of DNA fragments of known length, called a *DNA ladder*, is loaded in the first well. DNA samples to be studied are loaded in the remaining wells, and an electric current is applied to the apparatus. Since DNA is negatively charged, the DNA molecules in the wells travel toward the opposite, positive end of the gel. Smaller DNA fragments are able to move through the gel more easily and thus move faster than longer fragments. This causes the fragments to separate according to size as the procedure runs. Comparison to the DNA ladder provides an estimate of the separated fragments' sizes measured in kilobars (kB).

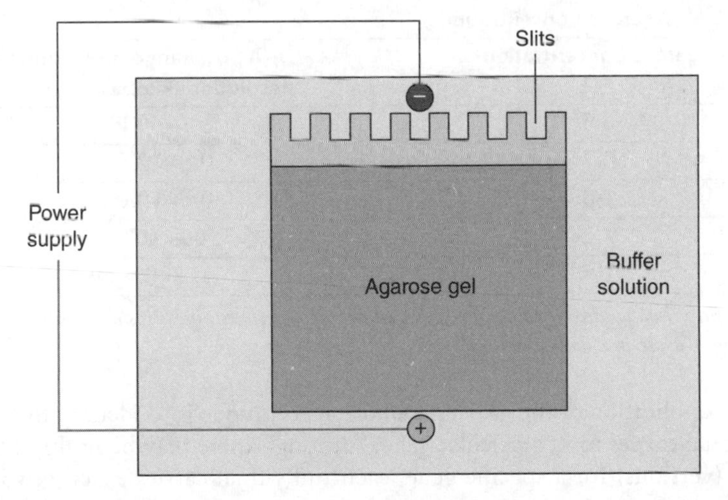

Figure 3.2

In addition to fragment size, several factors can affect the rate of migration of DNA fragments through the agarose gel. Table 3.4 provides a summary of the effects of agarose gel concentration and voltage of the electric current.

TABLE 3.4 Factors Affecting Fragment Migration

Factor	Variations	Effect on Fragment Migration
Agarose gel concentration		
(0.5–2%)	Low	Sharper resolution of larger DNA fragments Longer run time (may be days) Possibly weak/fragile gel
	High	Sharper resolution of smaller DNA fragments Shorter run time Possibly brittle gel
Voltage of electric current		
(0.25–7 V/cm)	Low	Sharper resolution of larger DNA fragments Longer run time (may be days) Possibility of DNA fragments <1 kb diffusing horizontally through gel
	High	Shorter run time Possible smearing of DNA fragments >10 kb Possibility of overheating gel, causing low-concentration gels to melt

Table 3.5 identifies the agarose gel concentration needed for optimum resolution of DNA fragments within various size ranges.

TABLE 3.5 Agarose Concentrations

Agarose Concentration (%)	DNA Size Range for Optimum Resolution (kilobases)
0.5	1–30 kb
0.7	0.8–12.0 kb
1.0	0.5–10.0 kb
1.2	0.4–0.7 kb
1.5	0.2–0.5 kb

Source: http://www.idtdna.com/pages/decoded/decoded-articles/pipet-tips/decoded/2011/06/17/running-agarose-and-polyacrylamide-gels.

One application of the gel electrophoresis technique is to identify the *alleles* an individual carries for a particular gene. Although there may be multiple possible alleles (versions) for a specific gene, each individual carries exactly two copies, one from each parent. When subjected to electrophoresis, each allele separates out into a distinct band, allowing that individual's pair of alleles to be identified. A single darker band in a particular lane indicates two copies of the same allele.

Figure 3.3 shows electrophoresis results for a gene with three possible alleles. Allele 2 is known to contain extra bases as compared to Allele 1. Allele 3 is known to be missing bases as compared to Allele 1. DNA samples from 16 different individuals are loaded in Lanes A through P. The sizes of the known fragments in the DNA ladder are listed along the left.

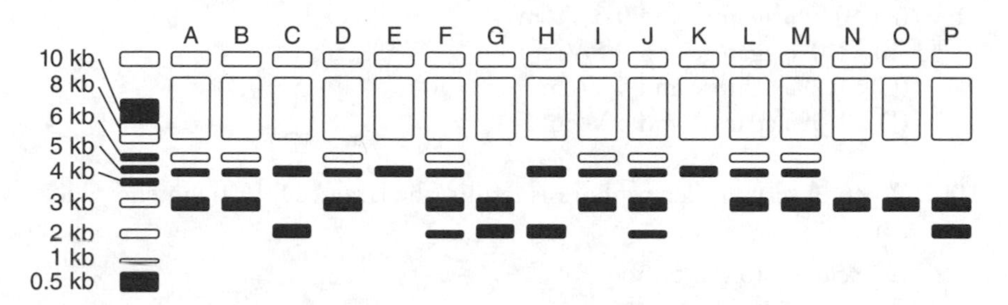

Figure 3.3

95. Applying an electric current to the electrophoresis apparatus causes the DNA fragments to travel:

(A) toward the wells.
(B) toward the cathode.
(C) away from the anode.
(D) away from the cathode.

96. According to the passage, the role of a DNA ladder is to:

(A) propel the DNA fragments through the agarose gel.
(B) provide an approximation of a DNA fragment's size.
(C) identify the base sequence in a DNA fragment.
(D) determine the total number of bases in a DNA fragment.

97. Which of the following would be a disadvantage of running a gel electrophoresis at a voltage of 6 V/cm?

(A) Poor resolution of large DNA fragments
(B) A short total run time for the procedure
(C) A total run time of more than one day
(D) Brittleness of the agarose gel

98. According to Table 3.5, as the concentration of agarose increases, the range of DNA fragment sizes that can be resolved:

(A) increases.
(B) is constant.
(C) decreases.
(D) increases, then decreases.

99. Which combination of factors would provide the best results for DNA fragments of 0.5–0.7 kb?

(A) 0.7% agarose and 0.5 V/cm
(B) 1.0% agarose and 0.5 V/cm
(C) 1.2% agarose and 5 V/cm
(D) 1.5% agarose and 5 V/cm

100. When observing electrophoresis results, the largest DNA fragments will appear:

(A) closest to the cathode.
(B) closest to the anode.
(C) as the largest bands.
(D) as the smallest bands.

101. Based on the information in the passage, Allele 2 traveled through the agarose gel faster than:

(A) Allele 1 but slower than Allele 3.
(B) neither Allele 1 nor Allele 3.
(C) Allele 3 but at the same rate as Allele 1.
(D) both Alleles 1 and 3.

102. According to the passage, a single darker band, as seen in Lane E, most likely indicates an:

(A) error during the electrophoresis process.
(B) error when collecting the DNA sample.
(C) individual missing an allele due to mutation.
(D) individual with two copies of the same allele.

103. What is the approximate size of Allele 1?

(A) 3.0 kb
(B) 1.0 kb
(C) 8.0 kb
(D) 0.5 kb

104. What is the most common allele combination represented in the DNA samples shown in Figure 3.3?

(A) Two copies of Allele 2
(B) Allele 1 and Allele 2
(C) Two copies of Allele 1
(D) Allele 1 and Allele 3

105. Which of the following provides the best explanation for the result shown in Lane J?

(A) Individual J carries Alleles 1 and 3.
(B) Individual J carries Alleles 2 and 3.
(C) Lane J contains the DNA ladder.
(D) Sample J contains DNA from two individuals.

106. Which allele combination is not represented in the DNA samples shown in Figure 3.3?

(A) Allele 1 and Allele 3
(B) Allele 1 and Allele 2
(C) Two copies of Allele 3
(D) Allele 2 and Allele 3

107. Which agarose concentration was most likely used in the electrophoresis in Figure 3.3?

(A) 1.5%
(B) 2.0%
(C) 1.0%
(D) 1.2%

108. Which combination of factors would cause the slowest migration of DNA fragments?

(A) 0.5% agarose and 7 V/cm
(B) 2.0% agarose and 7 V/cm
(C) 2.0% agarose and 0.25 V/cm
(D) 0.5% agarose and 0.25 V/cm

Passage 9

A student wanted to test human reaction time to different stimuli to determine the conditions that cause the fastest reaction. The student conducted three experiments to test reaction time.

Experiment 1

The student used a computer program to record the time between the sounding of a tone and the student pressing the spacebar on the keyboard. This process was repeated 10 times per trial. The program then averaged the 10 response times to produce an average for the trial. The student conducted three trials using a tone length of 200 milliseconds (ms) and three trials with a tone length of 400 ms. Results are shown in Table 3.6.

TABLE 3.6 Experiment 1

Trial	Response Time (ms)	Tone Length (ms)
1	158 ms	200 ms
2	154 ms	200 ms
3	152 ms	200 ms
4	144 ms	400 ms
5	142 ms	400 ms
6	143 ms	400 ms

Experiment 2

The student then used the same computer program to record the time between the sounding of a tone *or* the appearance of an image on the screen and the student pressing the spacebar. This process was repeated 10 times per trial, with the computer again averaging the 10 response times for each trial. The student conducted three trials using the tone as the stimulus and three trials using the image. Each stimulus lasted for a duration of 400 ms. Results are shown in Table 3.7.

TABLE 3.7 Experiment 2

Trial	Response Time (ms)	Stimulus
1	145 ms	Auditory
2	142 ms	Auditory
3	142 ms	Auditory
4	193 ms	Visual
5	189 ms	Visual
6	188 ms	Visual

Experiment 3

The student repeated the previous experiment but alternated the stimulus (tone versus image) with each trial. Results are shown in Table 3.8.

TABLE 3.8 Experiment 3

Trial	Response Time (ms)	Stimulus
1	143 ms	Auditory
2	195 ms	Visual
3	152 ms	Auditory
4	199 ms	Visual
5	151 ms	Auditory
6	199 ms	Visual

109. In the three experiments, response time is measured as the time between:

(A) exposures to two consecutive stimuli.
(B) exposure to a stimulus and the subsequent response.
(C) the registering of two consecutive responses.
(D) the beginning and end of one trial.

110. The stimulus in Experiment 1 was the:

(A) sounding of a tone.
(B) appearance of a screen image.
(C) pressing of the spacebar.
(D) use of a computer program.

111. How do Experiments 2 and 3 differ?

(A) Experiments 2 and 3 used different stimuli to test response times.
(B) The length of exposure to the stimulus was greater in Experiment 2.
(C) Experiment 3 included more trials than Experiment 2.
(D) In Experiment 3, the type of stimulus was alternated with each trial.

112. Based on the data in Table 3.7, the sense of hearing is:

(A) twice as fast as sight.
(B) more complex than sight.
(C) not as readily testable as sight.
(D) processed quicker than sight.

113. A stimulus duration of 400 ms was used during which experiment(s)?

(A) Experiments 2 and 3 only
(B) Experiment 1 only
(C) Experiments 1 and 2 only
(D) Experiments 1, 2, and 3

114. The fastest reaction time occurred in response to:

(A) an auditory stimulus lasting 200 ms.
(B) an auditory stimulus lasting 400 ms.
(C) a visual stimulus lasting 400 ms.
(D) a visual stimulus lasting 200 ms.

115. Based on the data in Table 3.6, what is the relationship between reaction time and length of stimulus exposure?

(A) Lengthening the stimulus improves reaction time.
(B) A shorter stimulus produces the fastest reaction time.
(C) Stimulus length has no measurable effect on reaction time.
(D) A longer stimulus produces the slowest reaction time.

116. Scientists have found that it takes 20–40 ms for a visual signal to reach the brain. Based on the data in Experiments 2 and 3, how long can an auditory signal be expected to take to reach the brain?

(A) 25–45 ms
(B) 50–55 ms
(C) 8–10 ms
(D) 20–40 ms

117. Scientists have found that a specific response time range exists for each particular sense. Which of the following would be the range for auditory stimuli?

(A) 140–160 ms
(B) 180–200 ms
(C) 150–170 ms
(D) 125–145 ms

118. How many total responses were recorded during Experiment 2?

(A) 10
(B) 6
(C) 60
(D) 30

119. Which graph best represents the data collected during Experiment 3?

(A) (B)

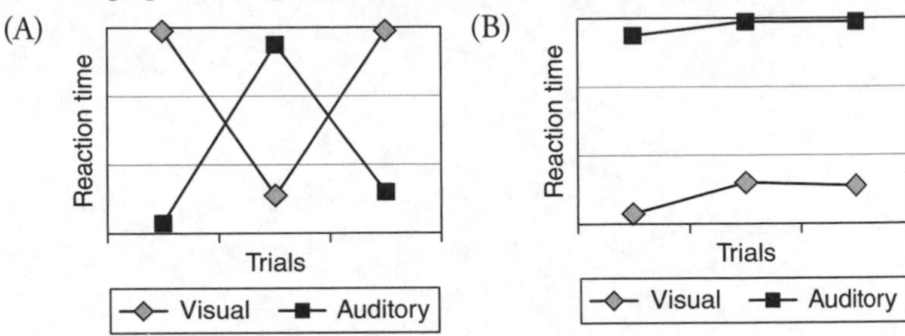

Figure 3.4 **Figure 3.5**

(C) (D)

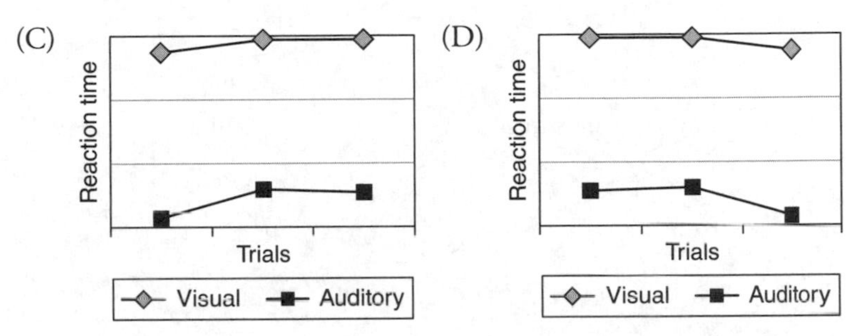

Figure 3.6 **Figure 3.7**

120. The student wants to test how varying the length of exposure to a visual stimulus affects response time. The best way to do this is to repeat:

(A) all three experiments using visual stimuli only.
(B) Experiment 1, replacing the tone with an image.
(C) Experiment 3, using a visual stimulus only.
(D) Experiment 2, using a stimulus duration of 200 ms.

121. What was the slowest auditory response time recorded during the three experiments?

(A) 199 ms
(B) 152 ms
(C) 142 ms
(D) 158 ms

122. Based on the data from the three experiments, what can be done to improve response time?

(A) Alternate exposure to two different stimuli.
(B) Decrease the duration of each exposure to a stimulus.
(C) Repeat exposure to the same stimulus.
(D) Increase the number of stimuli used at one time.

CHAPTER 4

Test 4

Passage 10

Organic molecules (carbohydrates, lipids, proteins, and nucleic acids) compose and are produced by living organisms. Scientists believe that simple organic molecules originally formed from inorganic molecules on primitive Earth. This step is considered a key precursor to the development of life on our planet. Two leading theories on the origin of the first organic molecules are described here.

Primordial Soup

The theory that organic molecules formed in the atmosphere of primitive Earth using energy from lightning is often called the "primordial soup theory." Evidence for this theory includes the Miller-Urey experiment, in which the conditions believed to exist in the primitive atmosphere were reproduced to create organic molecules.

The major components of the primitive atmosphere were believed to be methane (CH_4), ammonia (NH_3), hydrogen (H_2), and water (H_2O). These gases were put into a closed system and exposed to a continuous electrical charge to simulate lightning storms. After one week, samples taken from the apparatus contained a variety of organic compounds, including some amino acids (components of proteins). Figure 4.1 is a diagram of the apparatus used in the Miller-Urey experiment.

Hydrothermal Vents

The theory that organic molecules originally formed in the deep oceans using energy from inside the earth focuses on the existence of hydrothermal vents. Evidence for this theory includes the fact that ecosystems of diverse organisms have been found to exist around hydrothermal vents in the deep ocean. These ecosystems thrive without any energy input from the sun.

Organic molecules are only stable within a very narrow temperature range. Hydrothermal vents release hot (300°C) gases originating from inside the earth into the otherwise cold (4°C) water of the deep ocean. This release of gases causes a temperature gradient to exist around deep-sea vents. Scientists believe that within this temperature gradient exist the optimal conditions to support the formation of stable organic compounds. Figure 4.2 shows a diagram of the gradient produced by deep-sea vents.

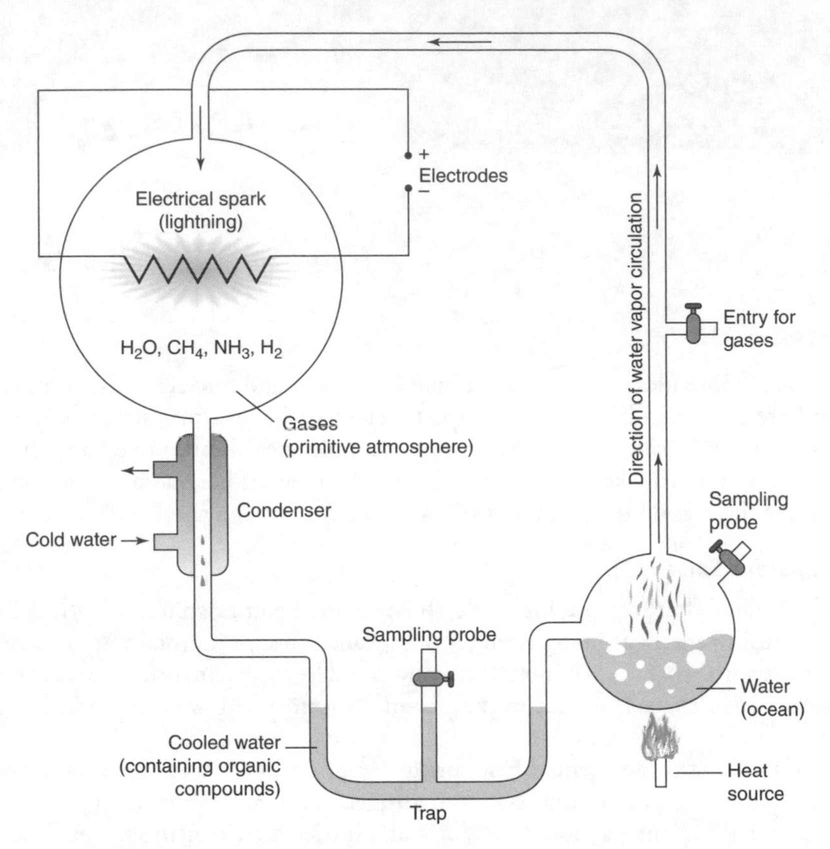

Figure 4.1

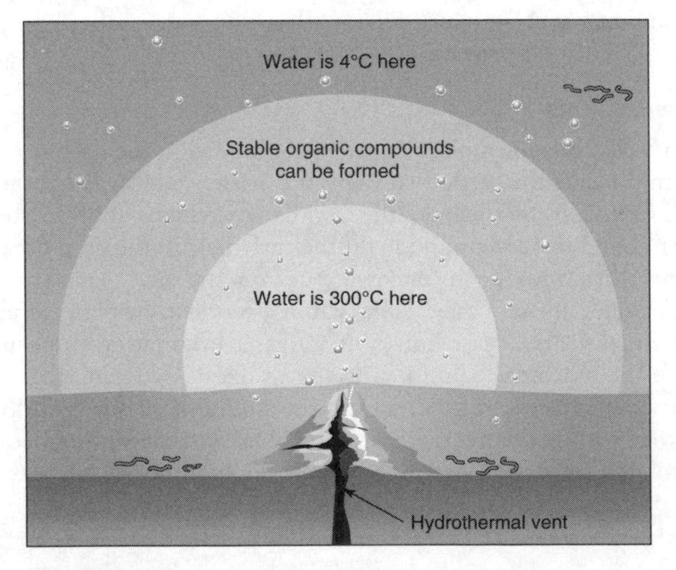

Figure 4.2

123. Which of the following is not an organic molecule?

(A) Carbohydrates
(B) Water
(C) Lipids
(D) Nucleic acids

124. Both theories on the origin of organic molecules are based on the assumption that those molecules:

(A) contain different atoms than inorganic molecules.
(B) only exist in the atmosphere and deep ocean.
(C) have not yet been produced in the laboratory.
(D) can be produced from inorganic molecules.

125. In Figure 4.1, the purpose of the heat source is to:

(A) produce water vapor for the simulated atmosphere.
(B) simulate hydrothermal vents in the deep ocean.
(C) reduce inorganic compounds to organic compounds.
(D) generate an electrical charge to stimulate the reaction.

126. Ammonia (NH_3) is an:

(A) inorganic compound.
(B) element.
(C) amino acid.
(D) organic compound.

127. In Figure 4.1, the reaction that produces organic molecules occurs in which part of the Miller-Urey apparatus?

(A) Condenser
(B) Large sphere
(C) Trap
(D) Small sphere

128. Based on the hydrothermal vents theory, which of the following would most likely be the optimal temperature range for organic molecule formation?

(A) Between 0°C and 4°C
(B) Higher than 300°C
(C) Lower than 300°C
(D) Between 4°C and 25°C

129. Which of the following statements would scientists supporting either theory most likely agree on?

(A) At least some organic compounds on Earth likely originated in meteorites from space.
(B) The production of amino acids requires the existence of a temperature gradient.
(C) The existence of water on Earth was essential to the original formation of organic compounds.
(D) A single method most likely produced the original versions of all organic molecules.

130. The specific source of energy used to form simple organic molecules is:

(A) addressed in the primordial soup theory only.
(B) not discussed in either of the two theories.
(C) a major difference between the two theories.
(D) the only similarity between the two theories.

131. According to the primordial soup theory, which of the following gases is not believed to have been a major component of the primitive atmosphere?

(A) Methane
(B) Hydrogen
(C) Water vapor
(D) Helium

132. The greatest limitation in the design of the Miller-Urey experiment is the:

(A) use of a condenser to cool water vapor.
(B) production of a variety of organic compounds.
(C) presence of continuous electrical sparks.
(D) recycling of water throughout the apparatus.

133. The hydrothermal vents theory states that organic molecules originally formed:

(A) inside the earth.
(B) in the earth's atmosphere.
(C) in the deep ocean.
(D) within volcanoes.

134. Scientists consider the outcome of the Miller-Urey experiment to be evidence:
(A) that refutes the primordial soup theory.
(B) in support of the primordial soup theory.
(C) that refutes both the primordial soup and the hydrothermal vents theories.
(D) in support of the hydrothermal vents theory.

135. According to the passage, temperature gradients exist in the deep ocean due to the:
(A) constant release of hot gases into cold water.
(B) decreased availability of sunlight at greater depths.
(C) existence of ecosystems made up of diverse organisms.
(D) reactions that produce organic molecules.

136. Which of the following is a key assumption of the primordial soup theory?
(A) Sunlight provided the energy needed to convert inorganic compounds to organic compounds.
(B) The composition of the primitive atmosphere was different than that of the current atmosphere.
(C) Amino acids can be produced from inorganic compounds in the laboratory.
(D) Organic compounds can only be produced by the reaction of other organic compounds.

Passage 11

As a liquid evaporates, the vapors on the surface of the liquid exert a *vapor pressure*. Vapor pressure varies with the liquid's temperature.

When vapor pressure equals the surrounding atmospheric pressure, boiling occurs. The *normal boiling point* of a liquid is defined as the temperature at which vapor pressure is equal to the standard atmospheric pressure of 760 mmHg (1 atm). If atmospheric pressure changes, a liquid's boiling point will also change.

Figure 4.3 illustrates the relationship between vapor pressure and temperature for four organic compounds belonging to the alkane group. The normal boiling point is indicated by a horizontal dashed line.

Organic compounds are composed of various functional groups attached to a hydrocarbon backbone. A *functional group* is a specific grouping of atoms that exhibits a characteristic set of properties. These properties remain consistent, regardless of the overall size of the compound.

Figure 4.4 compares the normal boiling points of organic compounds of increasing size for eight different functional groups, including the alkane group.

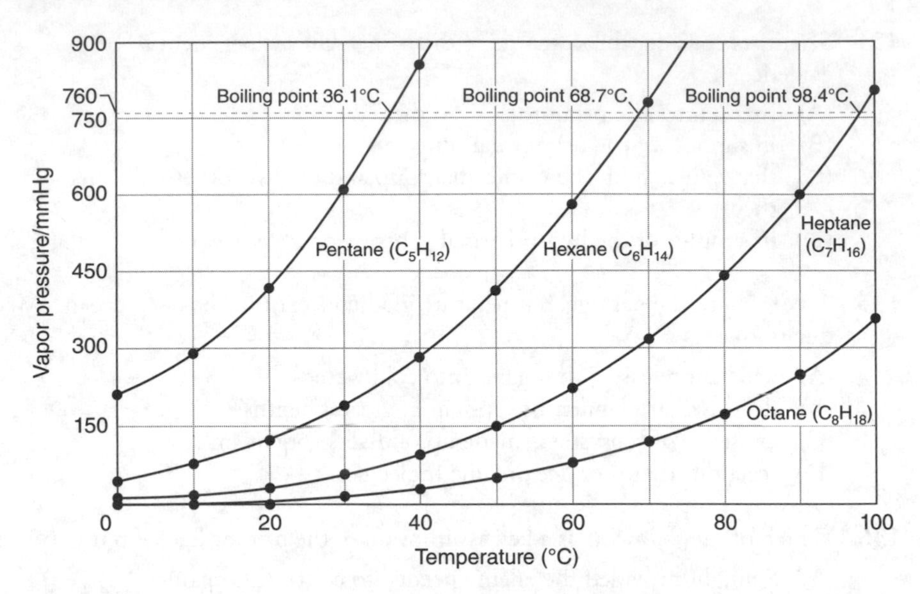

Figure 4.3

Source: http://wiki.chemprime.chemeddl.org.

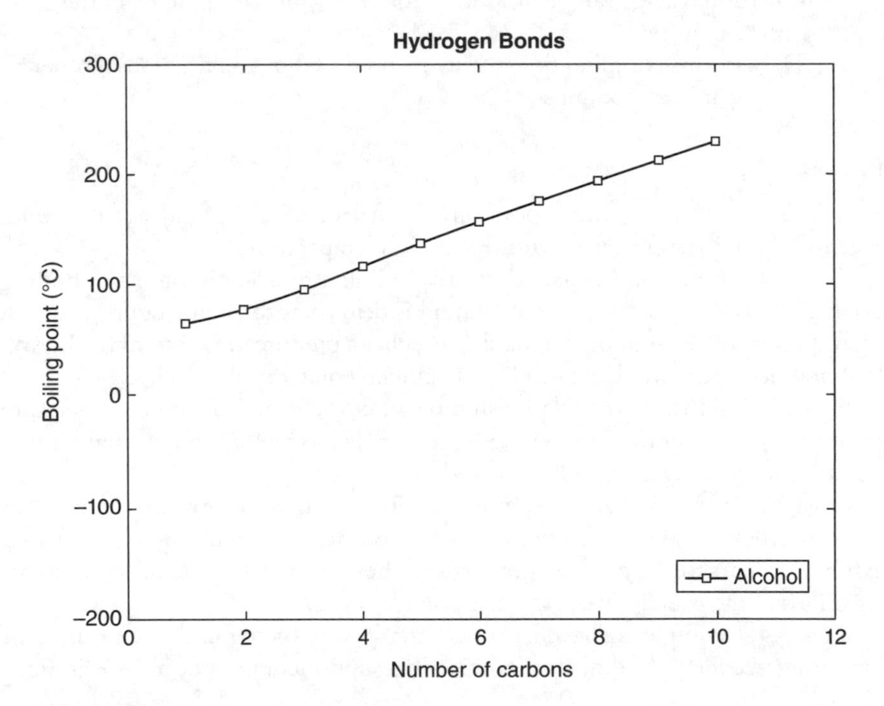

Figure 4.4

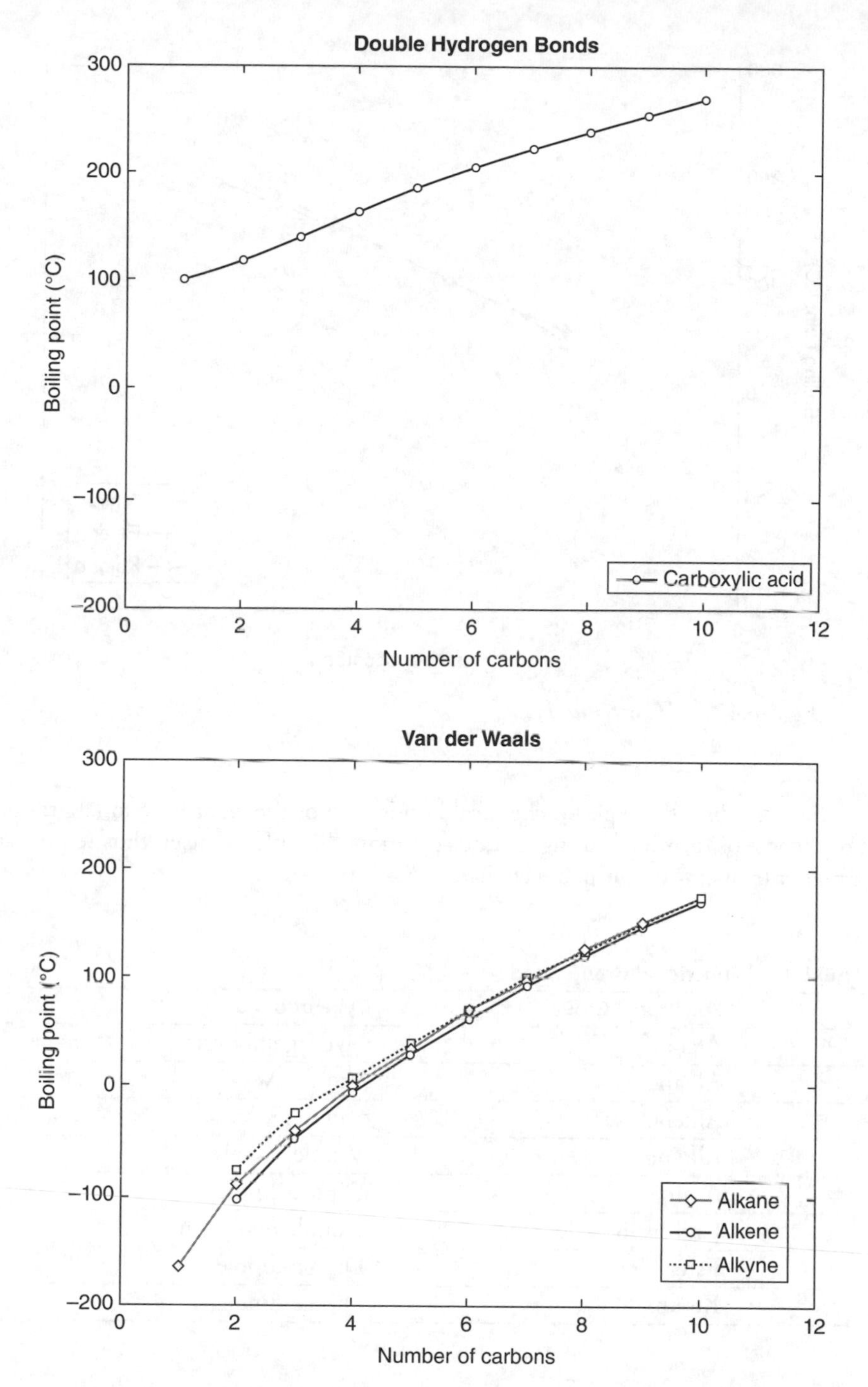

Figure 4.4 (***Continued***)

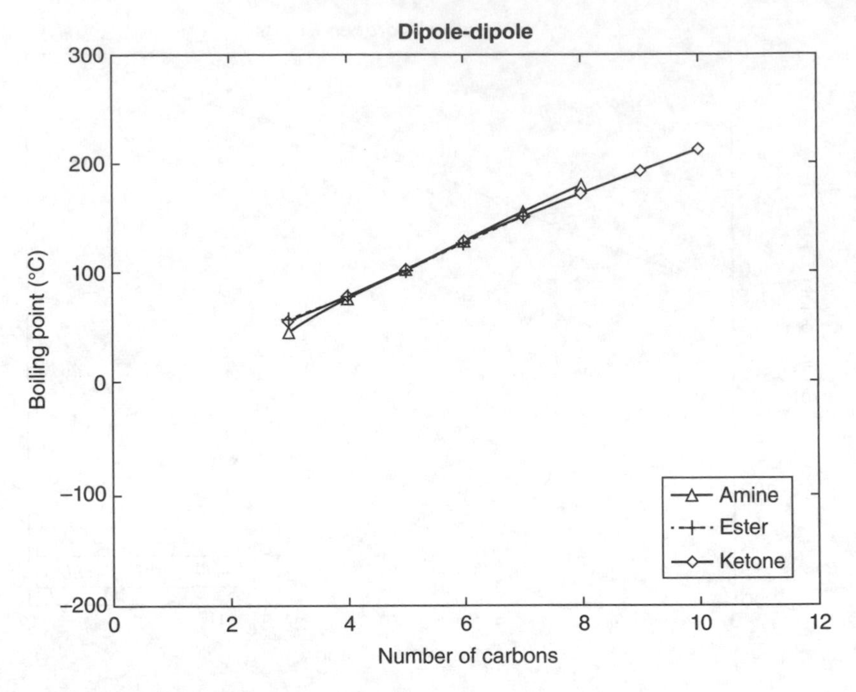

Figure 4.4 ***(Continued)***

Table 4.1 lists the types of chemical bonds each of the eight functional groups are capable of forming. Stronger bonds are more difficult to break, thus requiring a higher temperature for phase changes.

TABLE 4.1 Functional Group Bonds

Functional Group	Type of Bonds
Alcohol	Hydrogen bonds
Alkane	Van der Waals
Alkene	Van der Waals
Alkyne	Van der Waals
Amine	Dipole-dipole
Carboxylic acid	Double hydrogen bonds
Ester	Dipole-dipole
Ketone	Dipole-dipole

Table 4.2 lists characteristics of four common organic compounds with similar molecular weights. The temperatures listed represent the normal boiling point for each molecule.

TABLE 4.2 Molecular Weight

Molecule	Molecular Formula	Molecular Weight (g/mol)	Boiling Point (°C)
Propanoic acid	$C_3H_6O_2$	74	140
n-Butanol	$C_4H_{10}O$	74	117
Butanone	C_4H_8O	72	80
Pentane	C_5H_{12}	72	36

137. A compound's normal boiling point is the:

(A) minimum temperature at which the compound boils.
(B) average temperature at which the compound boils across all possible atmospheric pressures.
(C) maximum temperature at which the compound boils.
(D) temperature at which the compound boils under standard atmospheric pressure.

138. According to Figure 4.3, an organic compound will boil at a lower temperature if:

(A) vapor pressure increases.
(B) atmospheric pressure decreases.
(C) atmospheric and vapor pressures become unequal.
(D) vapor pressure is greater than atmospheric pressure.

139. At a vapor pressure of 50 mmHg, which alkane in Figure 4.3 would boil closest to 0°C?

(A) Heptane
(B) Hexane
(C) Pentane
(D) Octane

140. According to Figure 4.3, what vapor pressure will cause pentane's boiling point to be closest to 40°C?

(A) 760 mmHg
(B) 600 mmHg
(C) 400 mmHg
(D) 850 mmHg

141. What is the best approximation for the normal boiling point of octane in Figure 4.3?

(A) 126°C
(B) 100°C
(C) 145°C
(D) 98°C

142. According to Figure 4.4, the alkanes exhibit normal boiling points most similar to which other group?

(A) Alkynes
(B) Alcohols
(C) Carboxylic acids
(D) Amines

143. Based on the data in Figure 4.4, a 2-carbon alcohol would exhibit a normal boiling point closest to that of a:

(A) 3-carbon alkane.
(B) 9-carbon alkene.
(C) 4-carbon ketone.
(D) 2-carbon carboxylic acid.

144. Based on the data in Figure 4.4, which type of bond listed in Table 4.1 is the weakest?

(A) Dipole-dipole
(B) Double hydrogen
(C) Van der Waals
(D) Single hydrogen

145. Caproic acid is a carboxylic acid with a molecular formula of $C_6H_{12}O_2$. Which of the following temperatures is closest to the normal boiling point of caproic acid?

(A) 200°C
(B) 250°C
(C) 100°C
(D) 125°C

146. Based on the data in Figure 4.4, which of the following lists the bonds in Table 4.1 from the highest to the lowest boiling point required to break them?

(A) Van der Waals, dipole-dipole, single hydrogen, double hydrogen
(B) Double hydrogen, single hydrogen, dipole-dipole, Van der Waals
(C) Single hydrogen, double hydrogen, dipole-dipole, Van der Waals
(D) Dipole-dipole, Van der Waals, single hydrogen, double hydrogen

147. Which of the following generalizations about the relationship between an organic compound's molecular weight and its boiling point is best supported by the data in Table 4.2?

(A) The boiling point varies directly with molecular weight.
(B) As molecular weight increases, the boiling point decreases.
(C) As molecular weight decreases, the boiling point increases.
(D) The boiling point is not determined by molecular weight.

148. Based on the data in Figure 4.4, n-Butanol (see Table 4.2) most likely contains which functional group?

(A) Alcohol
(B) Ester
(C) Amine
(D) Alkyne

149. Based on the information in the passage, which of the following can be inferred about the type of bonds in an organic compound?

(A) Double hydrogen bonds are easier to break at high temperatures than single hydrogen bonds.
(B) Dipole-dipole bonds require the highest boiling point to break of all four types of bonds.
(C) Van der Waals bonds become easier to break as a compound's vapor pressure is increased.
(D) At the same vapor pressure, single hydrogen bonds require a higher boiling point to break than dipole-dipole bonds.

150. Which of the following generalizations is best supported by the data in Figures 4.3 and 4.4?

(A) Organic compounds containing the same number of carbon atoms have similar boiling points.
(B) The boiling point increases with the number of carbon atoms among organic compounds within the same group.
(C) The number of carbon atoms in an organic compound cannot be used to predict the compound's relative boiling point.
(D) The greater the number of carbon atoms in an organic compound, the lower that compound's boiling point is.

Passage 12

Shebay Park has been the site of ongoing population dynamics studies since the 1960s. Consisting of a group of isolated islands, the park provides ecologists with a unique, closed ecosystem in which to analyze the relationship between predator and prey populations. Figure 4.5 illustrates the food web for the Shebay Park ecosystem. In a food web, primary consumers are usually herbivores, feeding

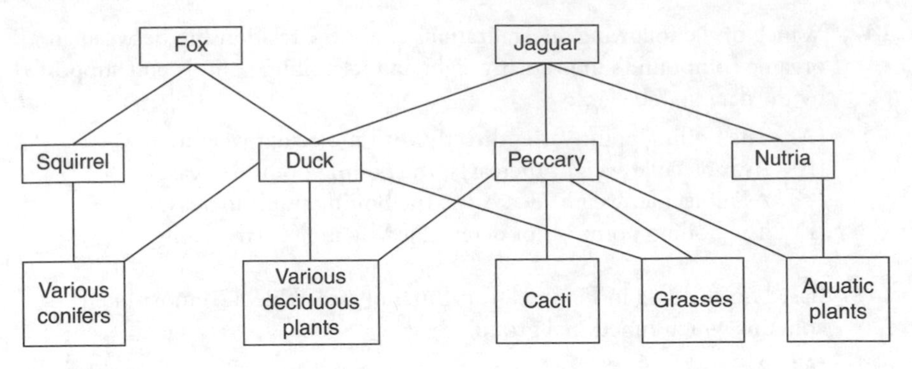

Figure 4.5

mainly on plants and fungi. Secondary consumers, on the other hand, feed mainly on primary consumers.

Ecological research in the park has focused mainly on the predator-prey relationship between the jaguar and peccary (a type of pig) populations. In addition to the typical selective pressures each species exerts on the other, scientists have observed specific events over the years that have affected population sizes. The inadvertent introduction of feline leukemia by humans in the late 1980s severely reduced the jaguar population. In 2004, the severest winter on record and an outbreak of ticks did the same to the peccary population. Figure 4.6 compares the annual population sizes for both species observed between 1968 and 2012.

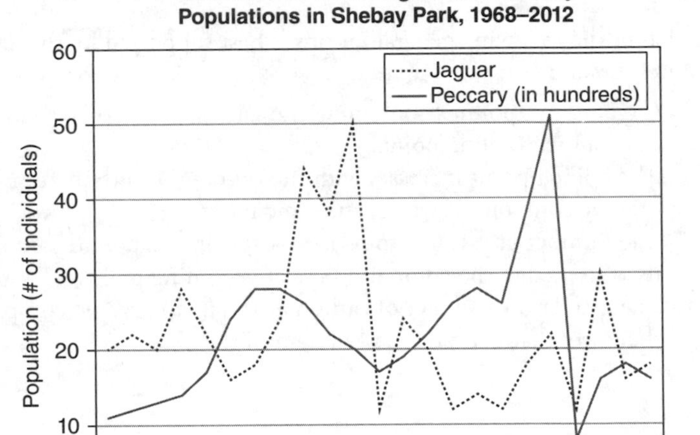

Figure 4.6

151. Shebay Park is considered a closed ecosystem because:
(A) organisms cannot easily migrate in from other ecosystems.
(B) population sizes within the ecosystem do not fluctuate.
(C) scientists have never had the opportunity to study the ecosystem.
(D) predator-prey is the only type of relationship that exists in the ecosystem.

152. According to the food web in Figure 4.5, peccary can be categorized as which type of consumer?
(A) Scavengers
(B) Herbivores
(C) Carnivores
(D) Omnivores

153. According to Figure 4.6, what has been the maximum size of the jaguar population since 1968?
(A) 20
(B) 60
(C) 50
(D) 10

154. The peccary population reached its smallest size in which year?
(A) 2006
(B) 2003
(C) 1995
(D) 1988

155. It can be inferred that the 13-year trend in the peccary population that began after 1990 was largely influenced by a sharp decline in:
(A) the jaguar population caused by disease.
(B) cactus growth caused by disease.
(C) the jaguar population during a severe winter.
(D) cactus growth during a severe winter.

156. According to Figure 4.5, how many secondary consumer species are present in the Shebay Park ecosystem?
(A) 0
(B) 1
(C) 6
(D) 2

157. Organisms that compete for many of the same resources within an ecosystem are said to occupy similar niches. Based on the information in Figure 4.5, which populations occupy a niche most similar to that of the peccary population?

(A) Nutria and squirrel
(B) Fox and jaguar
(C) Duck and fox
(D) Nutria and duck

158. Based on the data in Figure 4.6, a sharp decline in a population's size most commonly occurs in response to:

(A) a sharp increase in another population's size.
(B) an event that reduces individuals' immediate survival.
(C) a parallel decline in the size of other populations.
(D) an event that limits individuals' reproductive ability.

159. The increase in parasites may be partially responsible for the shift in:

(A) the peccary population after 2004.
(B) the peccary population before 2004.
(C) the jaguar population after 1990.
(D) the jaguar population before 1990.

160. Which of the following statements is best supported by the information in the passage?

(A) Predation is the single greatest factor affecting peccary population size.
(B) Food availability is the single greatest factor affecting peccary population size.
(C) Peccary population size varies independently of the predator population size.
(D) Predation is one of several factors that impact the size of the peccary population.

161. Based on Figure 4.5, which population is least likely to be affected by a change in the peccary population?

(A) Aquatic plants
(B) Squirrel
(C) Cacti
(D) Duck

CHAPTER 5

Test 5

Passage 13

An *invasive species* is a species that is not native to an ecosystem and whose introduction has harmful environmental, economic, and/or human health effects.

Eichhornia crassipes (water hyacinth) is an invasive species of floating aquatic weed found in freshwater waterways in tropical and temperate regions worldwide. It is highly tolerant of fluctuations in water level, nutrient availability, pH, and temperature. This allows it to grow rapidly and outcompete native aquatic plant species for resources. Dense floating mats of *E. crassipes* further alter aquatic communities by reducing dissolved oxygen levels and access to light. Decomposing matter from *E. crassipes* mats increases sediment deposition in waterways.

Ecological studies have shown that the growth of a plant can be influenced by competition with different species of neighboring plants. A group of scientists carried out the following studies to determine the effects on the growth of *E. crassipes* when paired with three other, more benign, aquatic weed species.

Study 1

Scientists collected growth data on *E. crassipes* mats in the Kagera River in Tanzania. Scientists marked off 1 square meter (m^2) sample areas containing *E. crassipes* alone and in combination with three other aquatic weeds common to the Kagera River.

To determine the effects of the other three weed species on *E. crassipes* growth, scientists analyzed five growth parameters. *Fresh weight* was determined by removing and immediately weighing 10 *E. crassipes* plants from each area. *Plant height* was measured from the base of the plant to the tip of the tallest leaf. The total number of *E. crassipes* plants within a sample area was recorded as *plant density,* which was then multiplied by fresh weight to determine *total biomass*. The number of leaves per plant was also recorded. Table 5.1 lists the averages for each growth parameter for *E. crassipes* growing alone and in combination with the three other aquatic weed species.

TABLE 5.1 Kagera River Data

Weed Combination	Fresh Weight (g)	Plant Height (cm)	Leaves per Plant	Plant Density (per m²)	Total Biomass (per m²)
E. crassipes	652.7	38.4	9.3	51.7	30.98
E. crassipes + *Commelina* sp.	452.2	33.7	9	38.9	16.09
E. crassipes + *Justicia* sp.	320.2	26.5	8.6	37.9	12.3
E. crassipes + *V. cupsidata*	332.7	21.8	8.6	26	7.77
E. crassipes + *Commelina* sp. + *Justicia* sp. + *V. cupsidata*	342	29.2	9	28.3	9.50

Source: http://www.academicjournals.org/ijbc/fulltext/2011/August /Katagira%20et%20al.htm.

Study 2

Scientists transplanted young *E. crassipes, Commelina* sp., *Justicia* sp., and *V. cupsidata* plants from the Kagera River to a greenhouse. In the greenhouse, *E. crassipes* potted alone and in combination with the other three weed species were allowed to grow in water from the Kagera River for four months. At the end of the four-month growth period, the parameters of fresh weight, plant height, and leaves per plant were all determined by the same methods used in Study 1.

Table 5.2 lists the averages for each growth parameter for *E. crassipes* growing in the greenhouse alone and in combination with the other aquatic weed species.

TABLE 5.2 Greenhouse Experiment Data

Weed Combination	Fresh Weight (g)	Plant Height (cm)	Leaves per Plant
E. crassipes	180.80	8.69	10.54
E. crassipes + *Commelina* sp.	129.08	8.80	10.70
E. crassipes + *Justicia* sp.	151.66	8.88	9.90
E. crassipes + *V. cupsidata*	98.19	8.24	9.75

Source: http://www.academicjournals.org/ijbc/fulltext/2011/August /Katagira%20et%20al.htm.

162. According to the passage, species identified as invasive are always:

(A) aggressively growing plants.
(B) disruptive to an ecosystem.
(C) introduced by humans.
(D) economically profitable.

163. According to the passage, water hyacinths upset freshwater ecosystems by doing all of the following except:

(A) increasing sediment deposition in waterways.
(B) outcompeting native plants for resources.
(C) altering the pH of aquatic environments.
(D) limiting aquatic organisms' access to sunlight.

164. Which weed combination was tested in Study 1 but not Study 2?

(A) Water hyacinth alone
(B) All four aquatic weeds together
(C) Water hyacinth with *V. cupsidata*
(D) Water hyacinth with *Justicia* sp.

165. In Study 1, plant density was measured as:

(A) the total number of *E. crassipes* plants in 1 m^2.
(B) the total number of weed plants in 1 m^2.
(C) fresh weight divided by water volume in 1 m^2.
(D) fresh weight divided by plant volume in 1 m^2.

166. Which weed combination serves as the control group in Study 1?

(A) *E. crassipes* with *Justicia* sp.
(B) *E. crassipes* with all three other weeds
(C) *E. crassipes* with *V. cupsidata*
(D) *E. crassipes* alone

167. Based on the data in Table 5.1, which weed exerts the least competitive pressure on *E. crassipes*?

(A) *Justicia* sp.
(B) *V. cupsidata*
(C) *Commelina* sp.
(D) The combination of all three weeds

168. In Table 5.2, *Commelina* sp. and *Justicia* sp. are both shown to have:

(A) a stronger effect on fresh weight than *V. cupsidata.*
(B) no effect on *E. crassipes* plant height.
(C) the same effect on fresh weight as *V. cupsidata.*
(D) a positive effect on *E. crassipes* plant height.

169. In Study 2, the water hyacinths grown alone exhibited a greater average:

(A) number of leaves than in Study 1.
(B) plant height than in Study 1.
(C) fresh weight than in Study 1.
(D) total biomass than in Study 1.

170. Total biomass was not included as a growth parameter in Table 5.2 because:

(A) plant density was not measured in Study 2.
(B) the fresh weight values recorded in Table 5.2 were too low.
(C) the plants used in Study 2 had no biomass.
(D) total biomass is not a good indicator of plant growth.

171. Which of the following statements is supported by the data collected in both studies?

(A) *V. cupsidata* has the most negative effect on water hyacinth growth.
(B) *Commelina* sp. has a positive effect on water hyacinth growth.
(C) Water hyacinth growth is not affected by the presence of other weed species.
(D) *Justicia* sp. has no effect on water hyacinth growth.

172. Based on the data in Table 5.2, the most significant impact of growing *E. crassipes* in combination with other weeds in a greenhouse environment appears to be the production of:

(A) shorter plants.
(B) lighter plants.
(C) fewer leaves per plant.
(D) fewer plants.

173. The greatest advantage of the experimental design in Study 2 is that scientists were able to:

(A) choose on which weed species to focus their observations.
(B) record data more frequently than could be done at the Kagera River.
(C) control for other environmental factors that may affect plant growth.
(D) obtain more precise measurements for each of the growth parameters.

174. According to Table 5.1, the presence of all three competitor weeds within the same square meter appears to have:

(A) a greater effect on *E. crassipes* fresh weight than the presence of any single competitor weed.
(B) an effect approximately equal to the sum of the effects of each single competitor weed on fresh weight.
(C) a lesser effect on *E. crassipes* fresh weight than the presence of any single competitor weed.
(D) an effect approximately equal to the mean of the effects of each single competitor weed on fresh weight.

175. Ecologists have found that introducing a competitor to an ecosystem is sometimes more effective in reducing an unwanted population than introducing a predator. Based on the results of this pair of studies, increasing the presence of which of the following species can be predicted to best reduce the water hyacinth population?

(A) *Justicia* sp.
(B) *V. cupsidata* and *Commelina* sp.
(C) *Commelina* sp. and *Justicia* sp.
(D) *V. cupsidata*

Passage 14

A roller-coaster car is often used as a model of energy transfer within a system. Resting at its starting point, the car has gravitational potential energy. As it moves along the track, the potential energy is converted to kinetic energy and then back to potential energy as the car approaches the roller coaster's ending point.

An object's gravitational potential energy can be calculated as the product of the object's mass, acceleration due to gravity, and the object's height above the ground ($PE_g = m \times g \times h$). In a frictionless system, the amount of potential energy at the beginning and end of the roller coaster would be equal, assuming that it arrives back at its starting position. However, friction between the car and the track causes *frictional dissipation* to transform some of the energy to sound and thermal energy. The amount of energy dissipated due to friction can be calculated as the product of the frictional force on an object and the distance traveled by the object ($F_f d$).

A group of students built a marble roller coaster out of foam pipe insulation tubing and tried to determine the conditions that would maximize the height it is able to reach. The students conducted two experiments to study the effects of gravitational potential energy and frictional dissipation on the marble.

Experiment 1

Figure 5.1 shows the initial setup for the marble roller coaster. A indicates the starting height (drop height) and C indicates the ending height (hill height) of the marble. B is the lowest point located halfway between A and C.

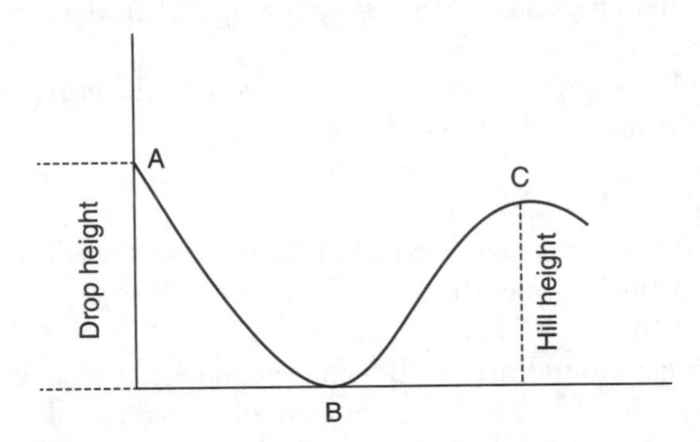

Figure 5.1

Students started with a drop height of 0.6 m and stretched the roller-coaster tubing out to a horizontal length of 1 m. They then varied the hill height until the marble was able to successfully reach the top of the hill without going over. To study the effects of the marble's initial gravitational potential energy, students conducted three more trials using different drop heights. Table 5.3 shows the results for each trial.

TABLE 5.3 Experiment 1

Trial	Drop Height (m)	Hill Height (m)
1	0.6	0.52
2	0.3	0.25
3	0.9	0.78
4	1.2	1.06

Experiment 2

Students started with a drop height of 1.2 m and stretched the roller-coaster tubing out to a horizontal length of 1.0 m. Students then varied the hill height until the marble was able to reach the top of the hill successfully without going over. To study the effects of frictional dissipation, students conducted two more trials using different horizontal track lengths. Table 5.4 shows the results for each trial.

TABLE 5.4 Experiment 2

Trial	Drop Height (m)	Horizontal Distance (m)	Hill Height (m)
1	1.2	1.0	1.06
2	1.2	0.5	1.15
3	1.2	1.5	0.97

176. When determining the gravitational potential energy of various objects on Earth, which variable would be considered a constant?

(A) h
(B) m
(C) PE_g
(D) g

177. According to the formula provided in the passage, doubling the height of an object should:

(A) double that object's potential energy.
(B) half that object's mass.
(C) double that object's mass.
(D) halve that object's potential energy.

178. In a frictionless environment, with a drop height of 0.6 m (A in Figure 5.1), the marble should be able to reach a hill height (C in Figure 5.1) of:

(A) 1.0 m.
(B) 0.6 m.
(C) 1.2 m.
(D) 0.5 m.

179. In Experiment 1, students altered the drop height of the marble to test the effect of which of the following variables on hill height?

(A) Frictional dissipation
(B) Horizontal distance traveled
(C) Initial gravitational potential energy
(D) Mass of the marble

180. What was the maximum drop height used in either experiment?

(A) 0.6 m
(B) 1.2 m
(C) 1.15 m
(D) 1.5 m

181. Which of the following graphs best represents the relationship between drop height and hill height in Experiment 1?

(A)

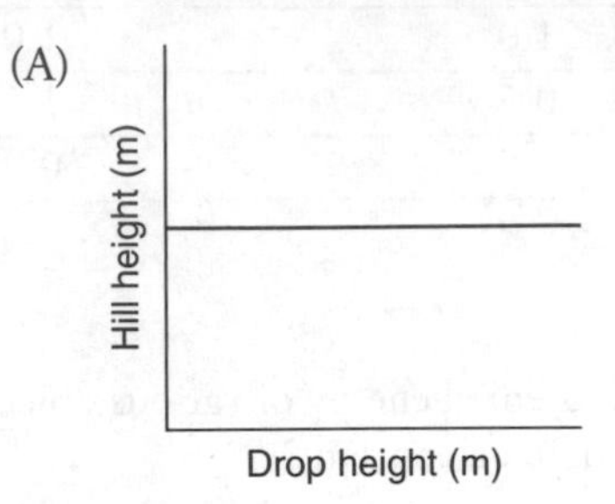

Figure 5.2

(B)

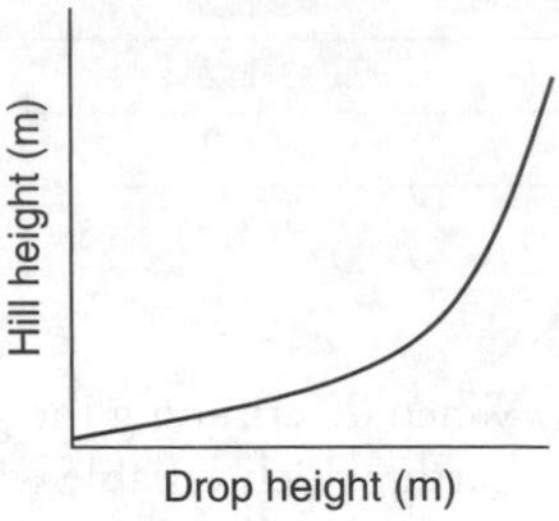

Hill height (m)
Drop height (m)

Figure 5.3

(C)

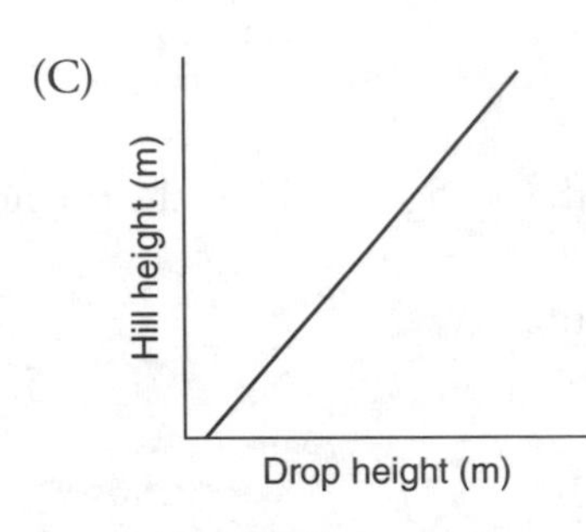

Figure 5.4

(D)

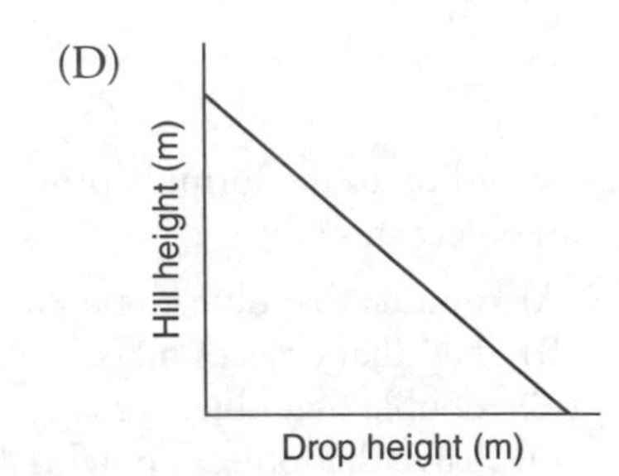

Figure 5.5

182. The data in Table 5.4 indicate that lengthening the roller coaster's track:

(A) causes the effects of frictional dissipation to increase.
(B) causes the effects of frictional dissipation to decrease, then increase.
(C) has no effect on the amount of frictional dissipation.
(D) causes the effects of frictional dissipation to decrease.

183. What was the smallest hill height recorded by the students in Experiment 2?

(A) 0.5 m
(B) 1.2 m
(C) 1.06 m
(D) 0.97 m

184. The students used drop height as the dependent variable in:

(A) Experiment 1 only.
(B) Experiment 2 only.
(C) both Experiments 1 and 2.
(D) neither Experiment 1 nor 2.

185. If the students were to carry out a third experiment to study the relationship between marble mass and hill height, how would the data table for this new experiment compare to Table 5.3?

(A) They would need to add an extra column between drop height and hill height for marble mass.
(B) They would need to replace the hill height column with a column for marble mass.
(C) They would need to add extra rows to the bottom of the table for additional trials.
(D) They would need to replace the horizontal distance column with a column for marble mass.

186. Based on the data for the two experiments, at which point in Figure 5.1 does the marble have the greatest gravitational potential energy?

(A) Point A
(B) Between points A and B
(C) Point C
(D) Point B

187. Which of the following is a similarity between Experiments 1 and 2?

(A) Both experiments began with an initial drop height of 0.6 m.
(B) The effect of the independent variable was studied by measuring hill height.
(C) Horizontal distance traveled was held constant in both experiments.
(D) The initial gravitational potential energy increased with each trial.

188. Which of the following energy transfers does not occur for the marble roller coaster in either Experiment 1 or Experiment 2?

(A) Mechanical energy to thermal energy
(B) Mechanical energy to sound
(C) Mechanical energy to chemical energy
(D) Potential energy to kinetic energy

189. In Experiment 2, a fourth trial using a horizontal distance of 1.75 m would most likely result in a hill height:

(A) greater than the hill height recorded in Trial 2.
(B) less than the hill height recorded in Trial 3.
(C) close to the hill height recorded in Trial 1.
(D) between the hill heights recorded in Trials 2 and 3.

Passage 15

The scientific classification of organisms provides information about the relative level of relatedness between species. Biologists use a hierarchical grouping system to classify organisms into various *taxa* (groups) based on shared physiological,

TABLE 5.5 Taxonomic Classification

	American Badger	Coyote	European Otter	Gray Wolf	Leopard
Kingdom	Animalia	Animalia	Animalia	Animalia	Animalia
Phylum	Chordata	Chordata	Chordata	Chordata	Chordata
Class	Mammalia	Mammalia	Mammalia	Mammalia	Mammalia
Order	Carnivora	Carnivora	Carnivora	Carnivora	Carnivora
Family	Mustelidae	Canidae	Mustelidae	Canidae	Felidae
Genus	*Taxidea*	*Canis*	*Lutra*	*Canis*	*Panthera*
Species	*taxus*	*latrans*	*lutra*	*lupis*	*pardus*

developmental, and genetic characteristics. Table 5.5 identifies the scientific classification of five common species.

Biologists use a *phylogenetic tree* to illustrate the evolutionary history of related species. In a typical tree, currently living species called *extant taxa* are listed along the right. Moving to the left, the point at which two or more extant taxa meet is called a *node.* A node indicates an *ancestral taxon*, or a common ancestor shared by the extant taxa.

Horizontal line length in a phylogenetic tree indicates relative *divergence time*, an estimation of how long ago the extant taxa are thought to have diverged into separate species. Figure 5.6 shows a phylogenetic tree of the species listed in Table 5.5.

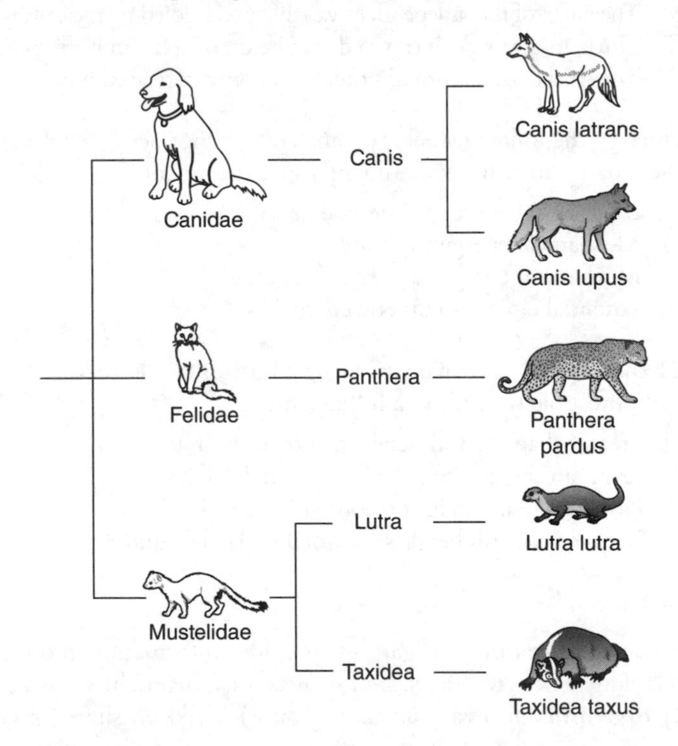

Figure 5.6

190. The phylogenetic tree in Figure 5.6 identifies evolutionary relationships between which type of organisms?

(A) Mammals
(B) Arthropods
(C) Invertebrates
(D) Amphibians

191. To which family does *Panthera pardus* belong?

(A) Mustelidae
(B) Canidae
(C) Felidae
(D) Carnivora

192. All organisms in Table 5.5 are members of the same:

(A) genus.
(B) order.
(C) species.
(D) family.

193. *Canids latrans* is the scientific name of which organism?

(A) American badger
(B) Gray wolf
(C) Leopard
(D) Coyote

194. According to Figure 5.6, the European otter is most closely related to which species?

(A) *Panthera pardus*
(B) *Canis lupus*
(C) *Taxidea taxus*
(D) *Canis latrans*

195. According to Figure 5.6, how many common ancestors does *Panthera pardus* share with *Taxidea taxus*?

(A) Three
(B) One
(C) Two
(D) Four

196. According to Figure 5.6, which pair of species have the most recent divergence time?

(A) *Lutra lutra* and *Panthera pardus*
(B) *Canis latrans* and *Lutra lutra*
(C) *Taxidea taxus* and *Lutra lutra*
(D) *Canis lupus* and *Canis latrans*

197. Based on the information in the passage and Figure 5.6, *Taxidea taxus* would be considered:

(A) an extant taxon.
(B) a node.
(C) an ancestral taxon.
(D) an order.

198. Which of the following inferences can be made about the seven-level classification system used in Table 5.5?

(A) Organisms classified in the same kingdom are classified in the same phylum.
(B) Organisms within the same class share a common kingdom and phylum.
(C) Organisms that share a common order cannot be classified in the same family.
(D) Organisms within the same family must share a common genus and species.

199. The scientific classification of the lynx is shown in Table 5.6. With which species would the lynx share the most recent ancestor?

TABLE 5.6 Lynx Classification

Kingdom	Animalia
Phylum	Chordata
Class	Mammalia
Order	Carnivora
Family	Felidae
Genus	*Lynx*
Species	*lynx*

(A) *Lutra lutra*
(B) *Canis latrans*
(C) *Panthera pardus*
(D) *Taxidea taxus*

200. How many taxonomic levels does the lynx have in common with the gray wolf?

(A) One
(B) None
(C) Two
(D) Four

201. Based on the information in Table 5.5, it can be predicted that the common ancestor shared by all five species belonged to which taxon?

(A) Mustelidae
(B) Carnivora
(C) Canidae
(D) Felidae

202. A *clade* is the taxonomic term for a grouping composed of all the descendants of a single ancestral taxon. According to Figure 5.6, which of the following groupings would not constitute a clade?

(A) *Canis latrans* and *Canis lupus*
(B) *Taxidea taxus* and *Lutra lutra*
(C) *Canis latrans*, *Canis lupus*, and *Lutra lutra*
(D) *Canis latrans*, *Canis lupus*, *Taxidea taxus*, and *Lutra lutra*

203. The wolverine (*Gulo gulo*) belongs to the family Mustelidae. Which of the following assumptions can be made about the wolverine?

(A) It is most closely related to the American badger.
(B) It shares the most genetic similarity with the European otter.
(C) It belongs to the same family as the gray wolf.
(D) It belongs to the same order as the coyote.

CHAPTER 6

Test 6

Passage 16

A gene is composed of a series of exon and intron segments. *Exons* are the coding regions of a gene, the segments that contain the instructions for building a protein. A gene's exons are connected by noncoding regions, or *introns.*

To build a protein, the cell must first transcribe the gene into messenger RNA (mRNA). Then a process called *RNA splicing* removes the noncoding introns and connects all of the exons to produce an mRNA transcript that can be used to build the protein.

Tropomyosins are a family of proteins that help maintain the cytoskeleton structure in all cells and support the contraction of muscle cells. In the late 1980s, a group of scientists discovered that the alpha-tropomyosin (α-TM) gene can code for several different tropomyosin proteins within different tissues of the same organism.

Figure 6.1 shows the structure of the seven mRNA transcripts identified as the result of the scientific study. In each transcript, each box represents an exon. Each transcript was found to be a product of the same α-TM gene.

Scientists continue to study the α-TM gene as a model of *alternative splicing*, in which mRNA transcripts containing different combinations of exons can lead to the production of different proteins. Figure 6.2 shows the structure of the α-TM gene, which is composed of 12 exons connected by 11 introns.

Each exon in a gene codes for a specific series of amino acids in the corresponding protein. The complete α-TM gene codes for a protein composed of 284 total amino acids. Table 6.1 shows the series of amino acids coded by each of the 12 exons in the α-TM gene.

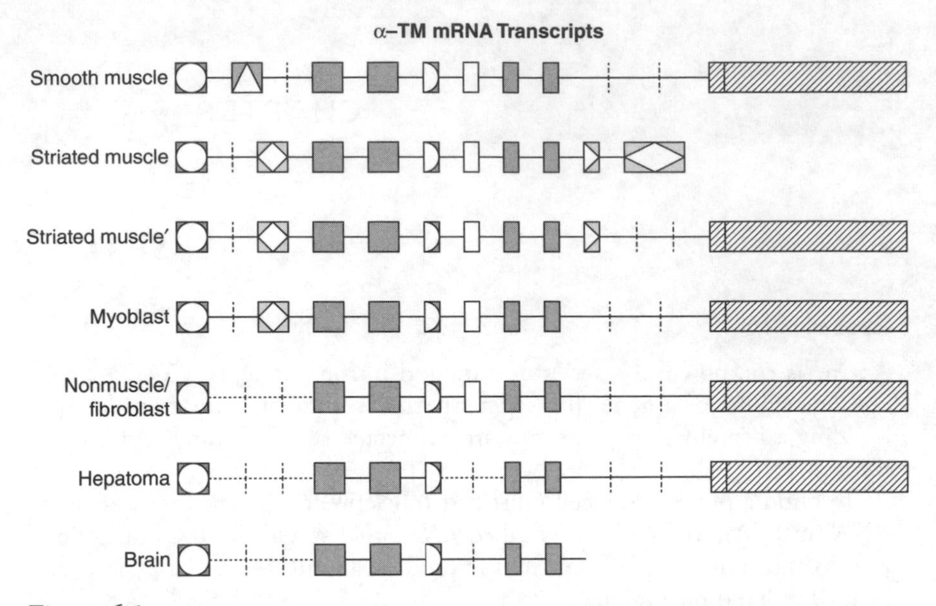

Figure 6.1
Source: http://www.bio.utexas.edu/research/tuckerlab/bright/phylo_3_10_04/index.html.

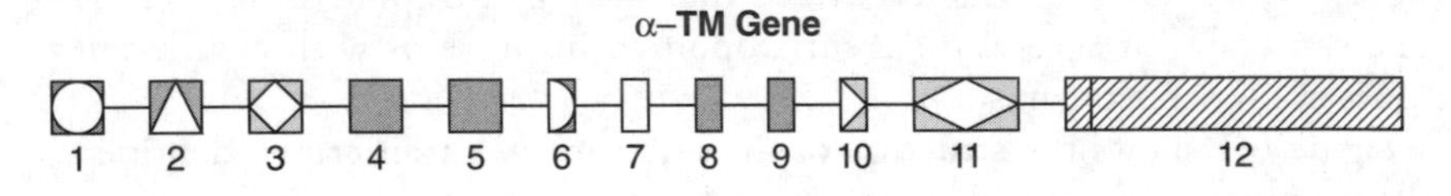

Figure 6.2
Source: http://www.bio.utexas.edu/research/tuckerlab/bright/phylo_3_10_04/index.html.

TABLE 6.1 Alpha-Tropomyosin Exon

Exon	Amino Acids
1	1–38
2	39–80
3	39–80
4	81–125
5	126–164
6	165–188
7	189–213
8	214–234
9	235–257
10	258–284
11	None
12	258–284

204. To produce a tropomyosin protein, which of the following steps must occur first?

(A) The introns are removed from the α-TM mRNA.
(B) Exons are alternatively spliced to code a specific tropomyosin.
(C) The α-TM gene is transcribed into mRNA.
(D) Amino acids are arranged based on the α-TM mRNA sequence.

205. In Figure 6.1, what is the maximum number of exons present in an mRNA transcript?

(A) 10
(B) 7
(C) 9
(D) 11

206. *Constitutive exons* are present in all mRNA transcripts of a gene and are thought to be integral in the proteins' basic structure. Which of the following exons appears to be constitutive?

(A) Exon 3
(B) Exon 7
(C) Exon 12
(D) Exon 4

207. *Alternatively spliced exons* (ASEs) are those that only appear in certain mRNA transcripts. Which of the following cell types appears to have the least number of ASEs?

(A) Myoblast
(B) Brain
(C) Nonmuscle/fibroblast
(D) Smooth muscle

208. Which exons do not appear in any of the same mRNA transcripts?

(A) Exons 10 and 12
(B) Exons 7 and 11
(C) Exons 2 and 3
(D) Exons 3 and 11

209. Two types of muscle tissues—skeletal and cardiac—are both striated. Based on Figure 6.1, how do the α-TM mRNA transcripts of skeletal and cardiac muscle tissues differ?

(A) One transcript contains a greater total number of exons.
(B) The exons present in one transcript are absent in the other.
(C) One contains Exon 2, while the other contains Exon 3.
(D) Each transcript contains a different final exon.

210. The total number of exons in the α-TM gene is:

(A) unknown.
(B) 12.
(C) variable.
(D) 11.

211. Based on the data in Table 6.1, which α-TM exon codes for the longest sequence of amino acids?

(A) Exon 8
(B) Exon 4
(C) Exon 11
(D) Exon 6

212. Which α-TM mRNA transcript is missing amino acids 258–284?

(A) Hepatoma
(B) Myoblast
(C) Smooth muscle
(D) Brain

213. Based on Table 6.1 and Figure 6.1, which mRNA transcript contains a repeated sequence of amino acids?

(A) Striated muscle
(B) Nonmuscle/fibroblast
(C) Smooth muscle
(D) Brain

214. A myoblast is an embryonic cell that can differentiate into a muscle cell. Based on Figure 6.1, which of the following happens to the α-TM mRNA transcript when a myoblast differentiates into a smooth muscle cell?

(A) Exon 10 is added.
(B) Exon 3 is replaced by Exon 2.
(C) Exon 12 is replaced by Exon 11.
(D) Exon 10 is removed.

215. *Untranslated regions* (UTRs) are sequences that exist at the beginning and end of every mRNA transcript. Instead of coding for amino acids, UTRs regulate the expression of the transcribed gene. In the α-TM mRNA, Exons 1 and 12 both contain UTRs. Based on the data in Table 6.1, which other exon contains a UTR?

(A) Exon 5
(B) Exon 8
(C) Exon 11
(D) Exon 3

216. The passage states that in addition to their function in all cells, tropomyosins also support contraction in muscle cells. It can be inferred that this extra function is related to which of the following sequences of amino acids?

(A) Amino acids 39–80
(B) Amino acids 81–125
(C) Amino acids 258–284
(D) Amino acids 1–38

217. A hepatoma is a tumor that forms within the liver. Based on Figure 6.1, it can be inferred that tumor formation may correlate to a loss of which exon?

(A) Exon 2
(B) Exon 10
(C) Exon 11
(D) Exon 7

Passage 17

Over the past several decades, scientists have seen a rapid decline in honeybee populations worldwide. In an effort to boost population sizes, the European Union recently instituted a temporary two-year ban on *neonicotinoids*, a class of pesticides thought to be harmful to honeybees.

Two scientists present their viewpoints regarding the value of instituting a similar ban in the United States.

Scientist 1

A short-term ban on the class of pesticides called *neonicotinoids* is a viable option that should seriously be considered by the United States. Studies have found neonicotinoid concentrations in pollen and nectar that can be lethal to pollinators. Although research has not identified a direct link between neonicotinoids and a reduction in honeybee populations, recent studies suggest that these pesticides may increase honeybees' susceptibility to parasites and diseases. The health of honeybee populations directly affects the agriculture industry and the overall ecosystem. Twenty-three percent of producer crops grown in the United States are pollinated by honeybees. Some crops, such as almonds, apples, onions, and carrots, are pollinated almost exclusively by honeybees. The reproductive rates of these crops vary directly with the availability of honeybees. Many of the plants that make up the base of the food web in the natural ecosystem also rely on these pollinators. Because the honeybee's role as pollinator is so pervasive, any measures that have the potential to support the health of honeybee populations should be taken.

Scientist 2

Honeybees are important pollinators for both natural ecosystems and the agriculture industry, and the health of their populations should be monitored closely. Instituting a ban on neonicotinoids, however, is unnecessary. Based on current research, the benefits of neonicotinoid use to the agriculture industry outweigh the threat to honeybee health. Though the exact causes are difficult to identify, researchers attribute the decrease in honeybee populations in recent years to weather, environmental stress, disease, and varroa mites. Environmental stressors include nectar and water that is scarce or of poor quality and exposure to pesticides, although researchers have found the latter to have the weakest correlation to honeybee loss of all stressors. Therefore, a ban on neonicotinoids will not be an effective approach for improving the health of honeybee populations. A more effective method should address varroa mites and disease, the greatest known threats to honeybee health.

218. According to the passage, neonicotinoids are a type of:

(A) parasite.
(B) pollinator.
(C) pesticide.
(D) pathogen.

219. According to Scientist 1, neonicotinoids:

(A) have been directly linked to declines in honeybee populations.
(B) may affect honeybees by increasing their vulnerability to parasites.
(C) provide agricultural benefits that outweigh the risk to honeybees.
(D) are the greatest threat to honeybee health in the United States.

220. Scientist 1 identifies all of the following crops as being highly dependent on pollination by honeybees except:

(A) cherries.
(B) almonds.
(C) carrots.
(D) apples.

221. The major difference between the two scientists' viewpoints is that:

(A) Scientist 1 believes honeybee populations should be saved, while Scientist 2 believes humans should not interfere with honeybee populations.
(B) Scientist 1 believes all threats to honeybee health should be addressed, while Scientist 2 believes that efforts should focus on the greatest threats to these populations.
(C) Scientist 1 believes honeybee populations are declining in the United States, while Scientist 2 believes that honeybee populations are stable.
(D) Scientist 1 believes neonicotinoids are harmful to honeybees, while Scientist 2 believes neonicotinoids do not pose any threat.

222. According to Scientist 1, which of the following graphs best represents the relationship between honeybees and producers in an ecosystem?

(A)

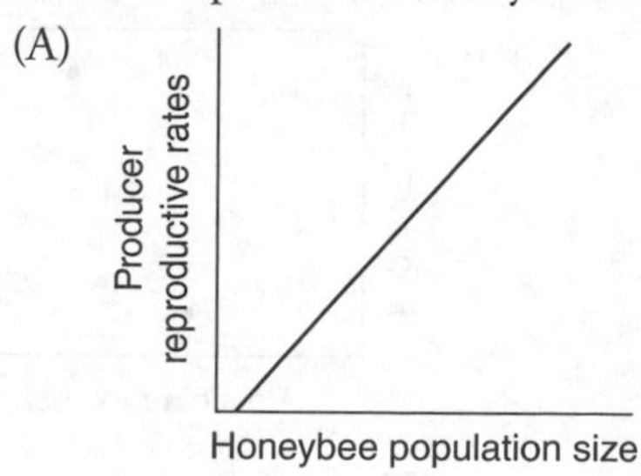

Figure 6.3

(B)

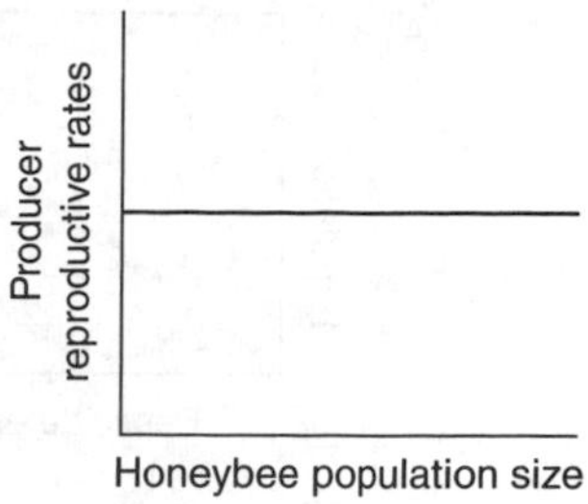

Figure 6.4

(C)

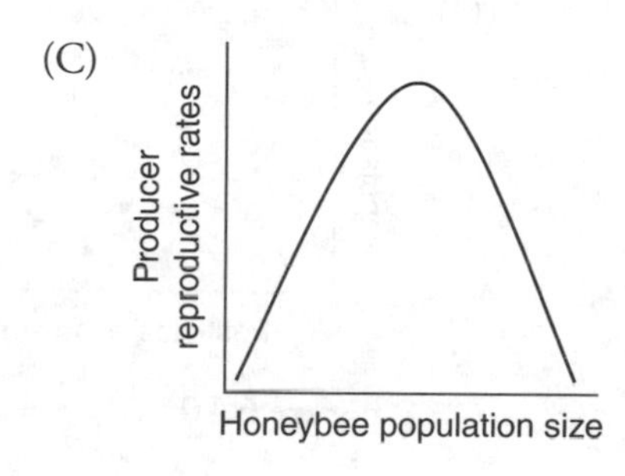

Figure 6.5

(D)

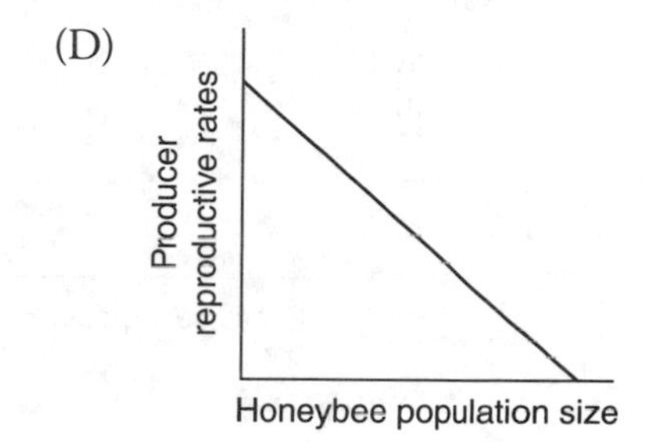

Figure 6.6

223. Which of the following does Scientist 2 identify as the greatest threats to honeybee populations in the United States?

(A) Varroa mites and disease
(B) Neonicotinoids and weather
(C) Disease and nectar quality
(D) Water and nectar scarcity

224. Which factor affecting honeybee health was discussed by Scientist 2 but not by Scientist 1?

(A) Pesticides
(B) Disease
(C) Parasites
(D) Water quality

225. It can be inferred that Scientist 1 believes honeybees' most important role in natural ecosystems is to:

(A) act as a host for varroa mites.
(B) provide a food source for birds.
(C) transfer pollen between plants.
(D) compete with other bee species.

226. According to Scientist 2, which graph best represents the relationship between neonicotinoid exposure and honeybee population?

(A)

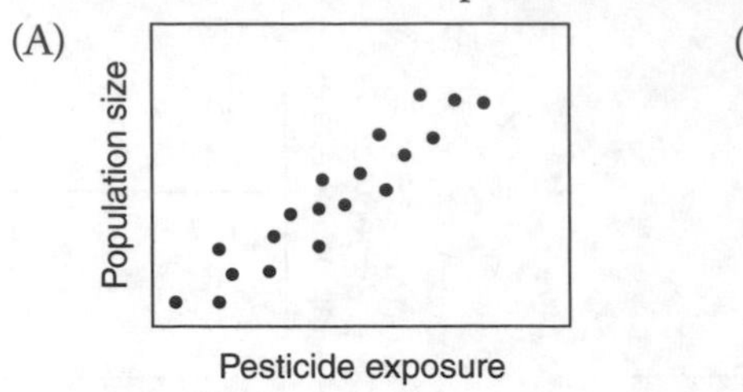

Figure 6.7

(B)

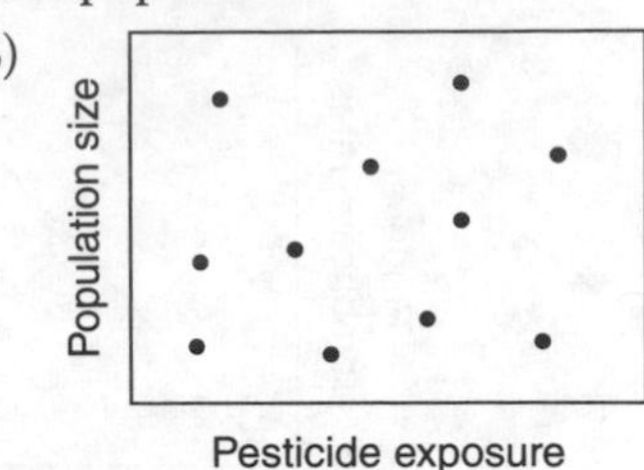

Figure 6.8

(C)

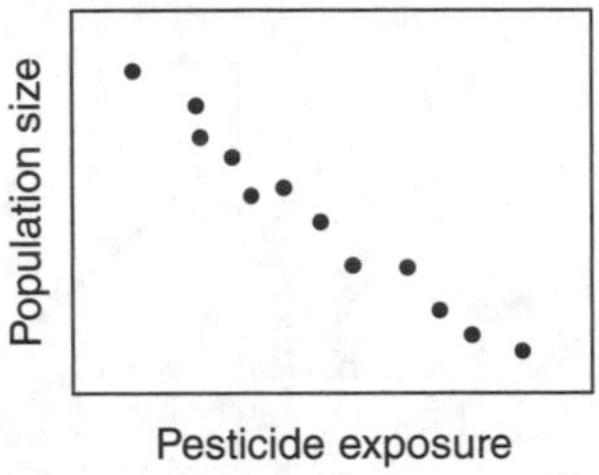

Population size

Pesticide exposure

Figure 6.9

(D)

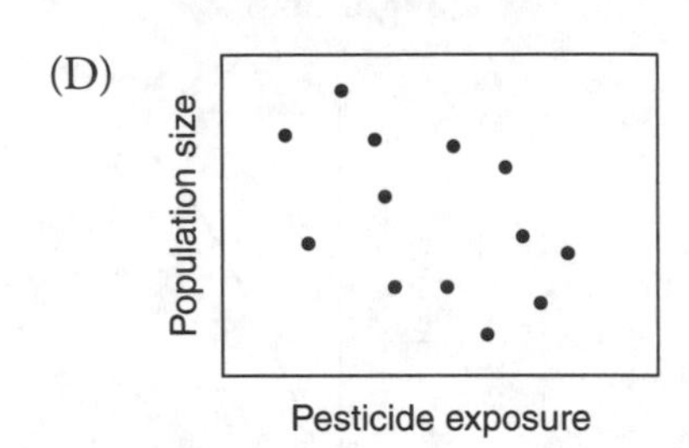

Figure 6.10

227. A doubling of the average honeybee population size in Europe over the next five years would support the opinion of:

(A) both scientists.
(B) neither scientist.
(C) Scientist 1.
(D) Scientist 2.

228. Based on the information in the passage, both scientists would support efforts to:

(A) institute a one-year ban on neonicotinoids in the United States.
(B) improve disease and parasite prevention in honeybee populations.
(C) reduce private consumer use of pesticides near honeybee habitats.
(D) monitor changes in the size of honeybee populations without interfering.

229. According to Scientist 1, approximately what fraction of the agriculture industry in the United States is dependent on honeybees?

(A) $\frac{1}{4}$

(B) $\frac{1}{20}$

(C) $\frac{1}{3}$

(D) $\frac{1}{23}$

230. If Scientist 2 is correct, which of the following trends is most likely to be seen if a ban on neonicotinoids is enacted in the United States?

(A) Honeybee populations will continue to decline at the preban rate.
(B) Honeybee populations will begin to increase at a rapid rate.
(C) Honeybee populations will continue to decline but at a slower rate.
(D) Honeybee populations will begin to increase at a moderate rate.

231. If Scientist 2 is correct, it can be inferred that honeybee health is most strongly affected by:

(A) seasonal conditions.
(B) resource availability.
(C) human interference.
(D) biotic factors.

Passage 18

Corals build the habitat that is the home for the fish and other marine species that live on the reef. The corals grow by creating aragonite forms of calcium carbonate cups in which the polyp sits. Figure 6.11 identifies the anatomy of a coral polyp.

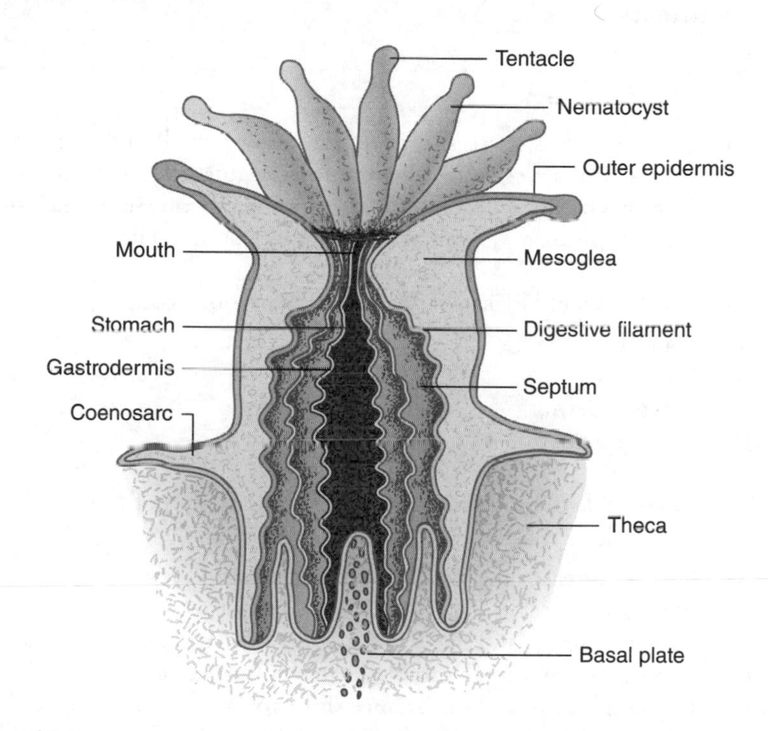

Figure 6.11

Source: http://oceanservice.noaa.gov/education/kits/corals/media/supp_coral01a.html.

Millions of photosynthetic algae, called zooxanthellae, reside inside polyp tissues. They serve as an energy source for corals as well as providing the coloration for which corals are known.

Current research indicates that increasingly acidic waters may be to blame for the decline in coral populations. Oceans absorb atmospheric carbon dioxide. Table 6.2 depicts changes to ocean chemistry and pH estimated using scientific models calculated from surface ocean measurement data.

TABLE 6.2 Ocean Chemistry and pH

	Preindustrial (1750)	Today (2013)	Projected (2100)
Atmospheric concentration of CO_2	280 ppm	380 ppm	560 ppm
Carbonic acid, H_2CO_3 (mol/kg)	9	13	19
Bicarbonate ion, HCO^{3-} (mol/kg)	1,768	1,867	1,976
Carbonate ion, CO_3^{2-} (mol/kg)	225	185	141
Total dissolved inorganic carbon (mol/kg)	2,003	2,065	2,136
Average pH of surface oceans	8.18	8.07	7.92
Calcite saturation	5.3	4.4	3.3
Aragonite saturation	3.4	2.8	2.1

Some coral become less successful at reproducing sexually in acidic waters. Studies also show links between ocean acidification and coral bleaching. Figure 6.12 summarizes the physiological responses of marine organisms to biological ocean acidification experiments done by various scientists.

232. Zooxanthellae would logically inhabit which part of a coral?

(A) Stomach
(B) Basal plate
(C) Outer epidermis
(D) Stinging nematocyst

233. The data in Table 6.2 indicate that as the concentration of carbon dioxide in the atmosphere rises:

(A) the pH decreases and the balance shifts toward bicarbonate instead of carbonate.
(B) the pH increases and the carbonate ion concentration increases.
(C) both the pH and the bicarbonate concentration decrease.
(D) the pH increases and the balance shifts toward carbonate instead of bicarbonate.

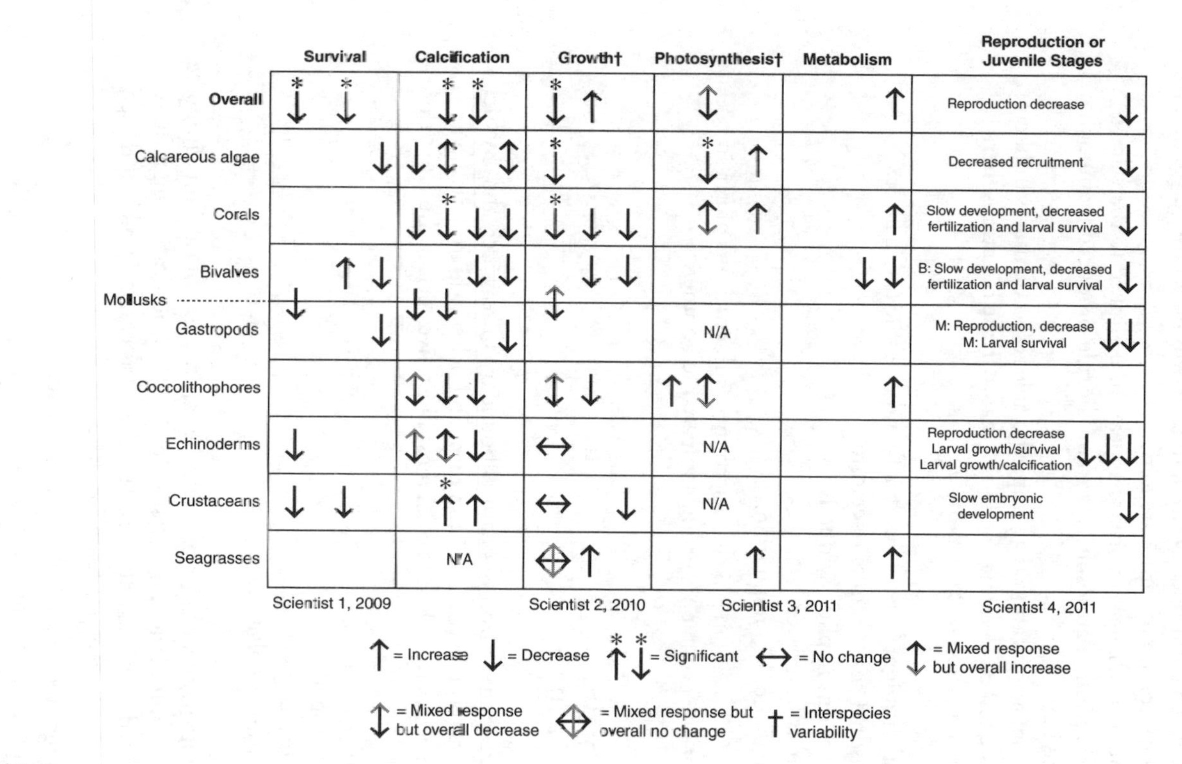

Figure 6.12

Source: Data adapted from "Recognising Ocean Acidification in Deep Time: An Evaluation of the Evidence for Acidification across the Triassic-Jurassic Boundary," Sarah E. Greene, Rowan C. Martindale, Kathleen A. Ritterbush, David J. Bottjer, Frank A. Corsetti, and William M. Berelson, Earth-Science Reviews, *volume 113 (1–2), copyright © 2012 by Elsevier.*

234. Based on the information in Table 6.2, what conclusions can be drawn about ocean chemistry?

(A) Future emissions of carbon dioxide are less likely to significantly impact ocean chemistry over time.
(B) Increased atmospheric carbon dioxide will have little impact on the concentration of carbonate ions.
(C) Chemical changes in oceans appears to be linked to the water's absorption of atmospheric carbon dioxide produced by human activities.
(D) Ocean acidification is an unpredictable response that is unlikely to be linked to human activities that increase the atmospheric concentration of carbon dioxide.

235. Factors that might impact the data found in Table 6.2 include:

I. seasonal changes in temperature.
II. variations in photosynthesis.
III. runoff from rivers.
IV. fluctuations in respiration.

Which of these is (are) likely to account for fluctuations of pH in ocean waters of one geographic region compared to another?

(A) II
(B) III
(C) I and IV
(D) II and III

236. The saturation horizon is a natural boundary in seawater, above which calcium carbonate ($CaCO_3$) can form and below which it dissolves. Which species from Figure 6.12 most likely lives below the saturation horizon?

(A) Corals
(B) Gastropods
(C) Crustaceans
(D) Calcareous algae

237. Calcifying organisms that produce the calcite form of calcium carbonate, such as foraminifera, can be less vulnerable to acidification than those constructed with aragonite structures, such as corals. Which of these provides a logical explanation for these findings?

(A) Aragonite is more soluble than calcite.
(B) Calcite is more soluble than aragonite.
(C) Aragonite saturation is farther from the surface of oceans.
(D) Calcite saturation is nearer to the surface of oceans.

238. Based on the information in Figure 6.12, decreased fertilization affects corals as well as:

(A) echinoderms.
(B) bivalves.
(C) crustaceans.
(D) calcareous algae.

239. In Figure 6.12, the most significant data with regard to the health of marine ecosystems is:

(A) the decline in coral calcification.
(B) the rise in crustacean calcification.
(C) the declining metabolism of bivalves.
(D) the decreased larval survival in gastropods.

240. According to Figure 6.12, which species appear to be most affected by ocean acidification?

(A) Echinoderms
(B) Gastropods
(C) Calcareous algae
(D) Crustaceans

241. Corals have several features that help them survive in the shallow ocean. Which part of a coral's anatomy may protect against fluctuating environmental changes such as temperature?

(A) Stomach
(B) Nematocysts
(C) Basal plate
(D) Outer epidermis

242. Some corals can reproduce in a variety of ways. Which of these methods would produce the most diverse offspring?

(A) Coral fragments regenerate to form new coral.
(B) Adult coral sprouts tiny buds to form new coral.
(C) Adult coral divides and both pieces grow new coral.
(D) Coral eggs join with coral sperm to form new coral.

243. Which of the following is most likely to result from declines in a coral polyp's zooxanthellae population?

(A) Bleaching
(B) Hyperpigmentation
(C) Increased thermal tolerance
(D) Accelerated growth of nematocysts

244. Scientists design an experiment in an attempt to predict the effect increasingly acidic seawater will have on coral reproduction. They may use the following in the experiment:

- Aquarium tanks
- Seawater
- Tap water
- Corals
- Carbon dioxide bubbles

Which experimental design will allow the scientists to investigate their hypothesis fairly and produce high-quality data for analysis?

(A) Use two tanks filled with seawater and corals. Add carbon dioxide bubbles to one tank.
(B) Use two tanks filled with tap water and corals. Add carbon dioxide bubbles to one tank.
(C) Use two tanks filled with corals. Add tap water to one tank and seawater to the other.
(D) Use three tanks filled with carbon dioxide bubbles. Add tap water and seawater to each tank.

245. Changes in the biological processes in the surface ocean water affect deeper portions of the ocean because:

(A) habitats at deeper levels depend on dissolved oxygen occurring at the surface.
(B) organisms living at lower ocean levels rely on products created by organisms at shallow levels.
(C) the pH of organisms in shallow waters is altered and becomes non-nutritious to deep-water organisms.
(D) the calcification of shallow-water organisms provides an additional layer of protection that prevents predation.

CHAPTER 7

Test 7

Passage 19

Lenses are made of transparent materials such as glass and plastic and are used in eyeglasses, cameras, and telescopes, as well as other applications. When light rays enter a curved lens from a distant object, the rays are bent into new angles. A *convex lens*, which is thicker in the middle, takes parallel light rays and converges them toward a common point called the focal point. The *focal length* is defined as the distance from the center of the lens to the point where the bent rays converge. A *concave lens*, on the other hand, is thinner in the middle and diverges parallel light rays as if they came from a point ahead of the lens (this point is one focal length from the lens). Both lenses are shown in Figure 7.1.

In conventional ray diagrams, the source of light (the object) is to the left of the lens and the rays move to the right through the lens. Light rays leave objects at various angles and are bent by the lens to form an image of the distant object. *Real images* are formed when actual light rays converge to a common point to the right of the lens. *Virtual images* are formed when the observer looks backward through the lens and sees an image on the same side of the lens as the object. Table 7.1 summarizes the images observed by these lenses.

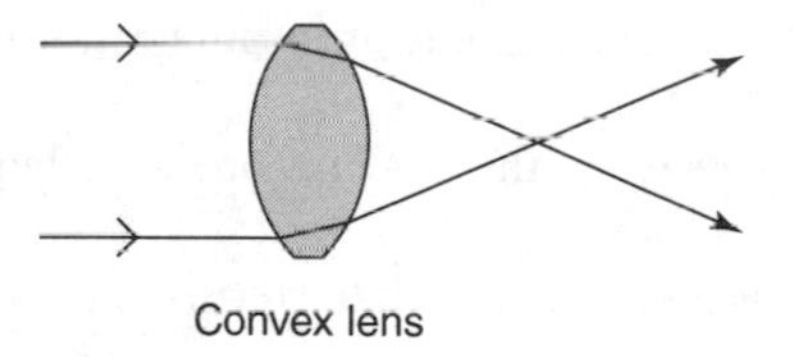

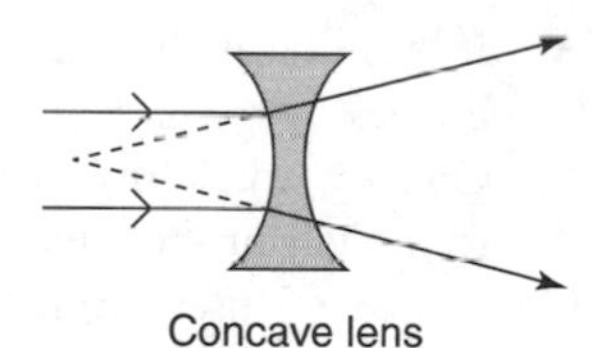

Figure 7.1

TABLE 7.1

Lens Type	Distance of Object to the Left of Lens	Image Description	Image Location
Convex	More than two focal lengths	Inverted, smaller	Between one and two focal lengths to the right of lens
Convex	Two focal lengths	Inverted, same size	Two focal lengths to the right of lens
Convex	Between one and two focal lengths	Inverted, larger	More than two focal lengths to the right of lens
Convex	One focal length	No image	No image
Convex	Within one focal length	Upright, larger	To the left of lens
Concave	Any position	Upright, smaller	To the left of lens

246. When the human eye views distant objects, the light rays go through a lens that is thicker in the middle. An image forms on the retina, which is the inner back surface of the eye. Which of the following best identifies the eye's lens and the characteristics of the image?

(A) Convex; real, inverted, and smaller
(B) Concave; virtual, upright, and smaller
(C) Convex; virtual, upright, and larger
(D) Concave; real, inverted, and smaller

247. According to the information provided in the passage and Table 7.1, which of the following statements is true?

(A) A lens that is thinner in the middle is capable of forming inverted images.
(B) Convex lenses form a real image when the object is one focal length to the left of the lens.
(C) Concave lenses can form images with many characteristics.
(D) All real images are inverted.

248. A particular convex lens in a camera has a focal length of 10 cm. If the object for the picture is 30 cm from the lens, what type of image will form on the film?

(A) A real, inverted, smaller image
(B) A real, inverted, larger image
(C) A real, inverted, same-size image
(D) A virtual, upright, smaller image

249. Rays of light from the distant sun reach the earth nearly parallel with each other. A child wishes to take these rays of light and use them to burn a piece of paper. What type of lens should the child use, and how far from the center of the lens should the paper be?

(A) A convex lens two focal lengths away from the paper
(B) A concave lens two focal lengths away from the paper
(C) A convex lens one focal length away from the paper
(D) A concave lens one focal length away from the paper

250. A slide projector uses a convex lens with a focal length of 120 mm. A small picture on a transparent slide is placed upside down in the projector, 125 mm in front of the lens. What will the image on the screen look like?

(A) Larger than the slide and right side up
(B) Larger than the slide and upside-down
(C) Same size as the slide and right side up
(D) Same size as the slide and upside-down

251. Eyeglass lenses may be used to correct both nearsighted and farsighted vision. Someone who is farsighted has difficulty seeing tiny objects that are close to the lens. As farsighted patients look through corrective lenses toward the object, they are able to see a larger image clearly on the same side of the lens as the object. What type of lens corrects farsighted vision, and what is the orientation of the image to that of the object?

(A) Convex; inverted
(B) Convex; upright
(C) Concave; inverted
(D) Concave; upright

252. A copy machine has a lens with a focal length of 30 cm. How far from the lens must a document (the object) be placed if the copy (the image) is to be exactly the same size?

(A) More than 60 cm from the lens
(B) 60 cm from the lens
(C) Between 30 and 60 cm from the lens
(D) 30 cm from the lens

253. A child looks through a lens at the distant trees, and the trees still look distant but appear smaller. What type of lens is the child looking through, and is the image upright or inverted?

(A) Convex; upright
(B) Convex; inverted
(C) Concave; upright
(D) Concave; inverted

254. A 15 cm focal length lens is used to focus a sharp image on a piece of paper that is 20 cm to the right of the lens. What information is known about the lens and the object?

(A) The lens is convex, and the object is 30 cm to the left of the lens.
(B) The lens is concave, and the object may be any distance from the lens.
(C) The lens is convex, and the object is between 15 and 30 cm to the left of the lens.
(D) The lens is convex, and the object is more than 30 cm to the left of the lens.

255. In Figure 7.2, Point F represents the focal point of the lens. If the candle is placed as shown in the figure, what type of image will be seen?

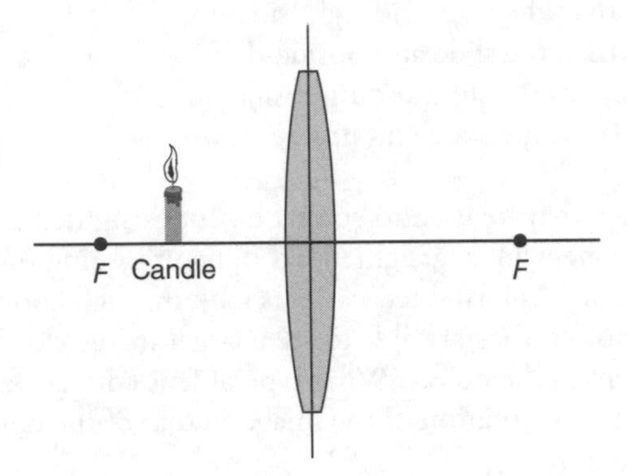

Figure 7.2

(A) A real, inverted, smaller image
(B) A real, inverted larger image
(C) A virtual, upright, larger image
(D) A virtual, upright, smaller image

Passage 20

In 1906, the Kaibab Plateau in northern Arizona was declared a federal game refuge by President Theodore Roosevelt. Before this time, the Kaibab was home to mule deer, cattle, sheep, and a variety of predators. The approximately 4,000 Rocky Mountain deer were an important source of food for the wolves, coyotes, bears, mountain lions, and bobcats that lived on the Kaibab and competed with sheep, horses, and cattle for the limited grass resources of the plateau.

When the game refuge was created, all deer hunting was banned in an attempt to protect the "finest deer herd in America." In 1907, the U.S. Forest Service began to exterminate the natural predators of the deer. With the deer freed from the checks and balances of predators, the population began to multiply. By the early 1920s, scientists estimated that there were as many as 100,000 deer on the plateau.

Sheep and cattle were also banned from the Kaibab. Signs of overgrazing were everywhere, and disease began to attack the crowded deer population. Hunting was reopened, but it was not enough to reduce the number of deer. Some estimate that as many as 60,000 deer starved to death in the winters of 1925 and 1926.

Two scientists exchange views about "The Kaibab Deer Incident: A Long-Persisting Myth."

Scientist A

The Kaibab Plateau should be a lesson to everyone about the disruption of the predator-prey relationship. This is a classic example of predator control hurting the very species that the wildlife biologists are attempting to help. If the predators had not been removed from the Kaibab Plateau, the deer population would have grown under normal conditions and would not have been subjected to the cruel fate of starvation and disease. This is a moral case that should be heeded by all biologists when considering predator control and presented to biology students in their studies of predator checks in population dynamics.

Scientist B

Predator removal is only a small part of the disaster on the Kaibab and has been grossly overdramatized. The deer population on the plateau grew rapidly because of the increase in food supply after the removal of competitive species. With no sheep and cattle to compete with for grazing, the environment could readily support more deer. The increased food supply allowed the population to grow quickly and to fall just as quickly due to the density-dependent factors of starvation and disease. In fact, data about the peak total number of deer on the plateau are unreliable, and there may have only been 30,000. The factors are more complex than early ecologists believed, but the lesson is still valuable.

256. Which of the following pieces of information would Scientist A use to support his claim?

(A) Before 1906, the Kaibab Plateau had already been overgrazed by the herbivores in the area.
(B) It is estimated that between 1907 and 1939, 816 mountain lions, 20 wolves, 7,388 coyotes, and more than 500 bobcats were killed.
(C) The U.S. Forest Service reduced the number of livestock grazing permits.
(D) In 1924, a committee formed to oversee the situation recommended that all livestock not owned by local residents be removed immediately.

257. Which of the following reflects evidence presented by both Scientist A and Scientist B about the deer situation on the Kaibab Plateau?

(A) Competition among herbivores was reduced due to restrictions on grazing.
(B) The food chain was disrupted when secondary consumers were reduced.
(C) Starvation and disease reduced the herd during the winters of 1925 and 1926.
(D) Human intervention in the predator population was the cause of the upsurge in the deer population.

258. Which statement would LEAST likely be attributed to Scientist B?

(A) "Data about the deer herd are unreliable and inconsistent, and the factors that may have led to an upsurge are hopelessly confounded."
(B) "Conclusions that have been made about the Kaibab are based on the maximum estimate and evolved by unjustified tampering with original data."
(C) "This is a classic example of how the effects of disruption of the predator-prey relationship can be seen plainly."
(D) "The reduction in sheep alone from 1889 to 1908 might have totaled 195,000."

259. The following statements have been made by biologists to describe the Kaibab Plateau situation. On which statement would Scientists A and B be likely to agree?

(A) "The plateau represents the unforeseen and disastrous possibilities of ignorant interference in natural communities."
(B) "The Kaibab is a classic example of what happens when people set out to protect prey from their "'enemies"' (sometimes only to preserve them for their human ones) by killing the predators."
(C) "Man is the most destructive predator alive."
(D) "This situation is a well-documented example of what can happen when predators are removed from an ecosystem."

260. The views of Scientist A:

(A) minimize the role of the bounty placed on predators.
(B) emphasize the lack of competition for resources.
(C) show a more balanced view of the problem by taking into account all factors that led to the increase in population.
(D) are likely to be used by someone trying to illustrate the dangers of removing a species from the food chain.

261. Scientist A would be most likely to support:

(A) the introduction of non-native species into an area where there are no natural predators.
(B) controlled hunting of predators to protect endangered species.
(C) future efforts to reorganize natural ecosystems through human intervention.
(D) the view that predators help preserve ecosystems.

262. Scientists A and B tend to agree on:

(A) the role of an increase in grass abundance in the increase of the deer population.
(B) the role that disease and starvation played in reducing the population.
(C) the role of predation in the increase of the deer population.
(D) the data to be used to represent the situation on the Kaibab.

263. Which of the following facts would support the view of Scientist B regarding the cause for rapid increase in the deer population?

(A) Coyotes were hunted in the thousands.
(B) Starvation and disease were rampant from 1924 to 1926.
(C) Hunters killed 674 deer in 1924.
(D) Sheep and cattle were banned during this time period.

264. Lethal reduction of midsized mammal predators that target duck nests is a method used to increase the duck population available for sport hunters. Which of the following statements would Scientist A most likely make regarding this practice?

(A) Hunting will keep the duck population from increasing unchecked and limit growth.
(B) The removal of mammals such as foxes and skunks will disrupt other areas of the food chain, such as the population of mice.
(C) The duck population will have greater nesting success as a result of reduced predatory concerns.
(D) Other waterfowl will enjoy the benefits of less predation.

265. On which of the following conclusions would Scientists A and B agree?

(A) Human intervention in natural ecosystems is a necessary step to protect populations.
(B) Caution should be taken when creating an ecological situation that favors a single species.
(C) Humans can make a change involving a single species with little or no effect on other species in the area.
(D) The Kaibab Plateau does not offer any lessons applicable to modern-day issues in ecology.

Passage 21

A group of students gathered data to determine the factors that affect the speed of a wave pulse as it travels down a spring. They studied springs with their coils stretched out (high tension) against springs with looser coils (low tension) to determine how changing the characteristics of the medium affect wave speed. They also studied the effect of wave amplitude on the speed of the wave. *Amplitude* is the size of the disturbance.

The students conducted slow-motion video analysis of a wave pulse traveling down the spring and graphed the total distance the pulse traveled versus the total travel time, as shown in Figure 7.3.

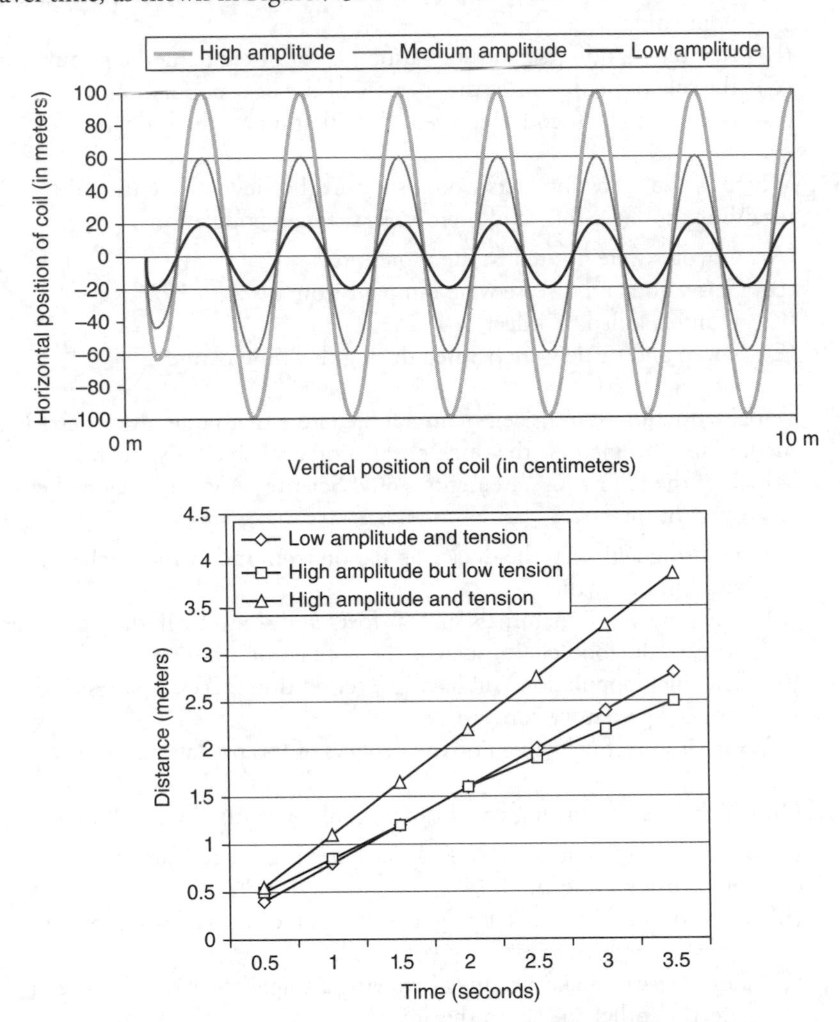

Figure 7.3

Source: U.S. Geological Survey and www.thinkquest.org. http://earthquake.usgs.gov/learn/glossary/?term=amplitude.

266. To properly determine the effect of wave amplitude on wave speed, the students must:

(A) make the spring tighter as they change the amplitude.
(B) keep the tension the same but change the amplitude.
(C) keep the amplitude and tension the same.
(D) change both the tension and the amplitude.

267. According to Figure 7.3, the average speed of the high-tension/high-amplitude wave is closest in value to:

(A) 0.50 m/s.
(B) 0.72 m/s.
(C) 0.91 m/s.
(D) 1.10 m/s.

268. According to Figure 7.3, the low-tension/low-amplitude wave:

(A) gained speed with time.
(B) lost speed with time.
(C) maintained a constant speed.
(D) accelerated at first and then slowed down.

269. According to Figure 7.3, what is the effect of the tension of the spring on wave speed?

(A) An increase in tension results in a greater wave speed.
(B) An increase in tension results in a smaller wave speed.
(C) Tension has no significant effect on wave speed.
(D) Tension increases the amplitude of the wave.

270. What is the effect of amplitude on wave speed?

(A) An increase in amplitude results in a greater wave speed.
(B) An increase in amplitude results in a smaller wave speed.
(C) Amplitude has no significant effect on wave speed.
(D) Amplitude increases the tension of the wave.

271. Because sound behaves like a wave, one can infer from the results in Figure 7.3 that loud sounds:

(A) travel at the same speed as soft sounds.
(B) travel faster than soft sounds.
(C) gain speed as they move through the air.
(D) lose speed as they move through the air.

272. It is believed that, along with other characteristics, light behaves like waves. From the results in Figure 7.3, one can infer that:

(A) the speed of light depends on the characteristics of the material it goes through.
(B) the speed of light depends on the brightness of the light.
(C) light loses speed as it moves away from its source.
(D) increased light frequency increases wave speed.

273. Water waves behave like spring waves in many ways. From the results in Figure 7.3, one can conclude that a water wave:

(A) gains speed as it travels.
(B) loses speed as it travels.
(C) gains speed at first before slowing down to rest.
(D) maintains a constant speed as it travels.

274. Assuming the low-tension/low-amplitude wave was able to keep moving, what is the approximate distance it would travel in 6.0 seconds?

(A) 0.77 m
(B) 1.6 m
(C) 4.8 m
(D) 6.4 m

275. In another experiment with springs, a large wave travels 10.2 m along a spring in 5.3 seconds. Approximately how much time would it take a small ripple to travel the same distance in the spring?

(A) 1.9 seconds
(B) 4.9 seconds
(C) 5.3 seconds
(D) 10.2 seconds

Passage 22

The *boiling point* of a liquid is commonly defined as the temperature at which the vapor pressure of the liquid is equal to the atmospheric pressure that surrounds the liquid. When a liquid is brought to a temperature at or above its boiling point, it quickly changes from the liquid phase to the gaseous phase. This can easily be observed in vapor bubbles that form in the liquid and rise to the top.

The boiling point of a substance depends on many different factors, such as the composition of the substance, its molar mass, and the atmospheric pressure surrounding it. The names, molar masses, and skeletal models of a variety of *alkanes* and three *alcohols* were identified and recorded in Table 7.2.

TABLE 7.2

	Name	Molar Mass (g/mol)	Skeletal Model
Alkanes	Pentane	72	
	Hexane	86	
	Heptane	100	
Alcohols	Butanol	74	OH
	Pentanol	88	OH
	Hexanol	102	

Figure 7.4 shows the boiling points of these six substances versus their molar mass. All temperatures are in degrees Celsius and have been recorded at a standard atmospheric pressure of 1 atm.

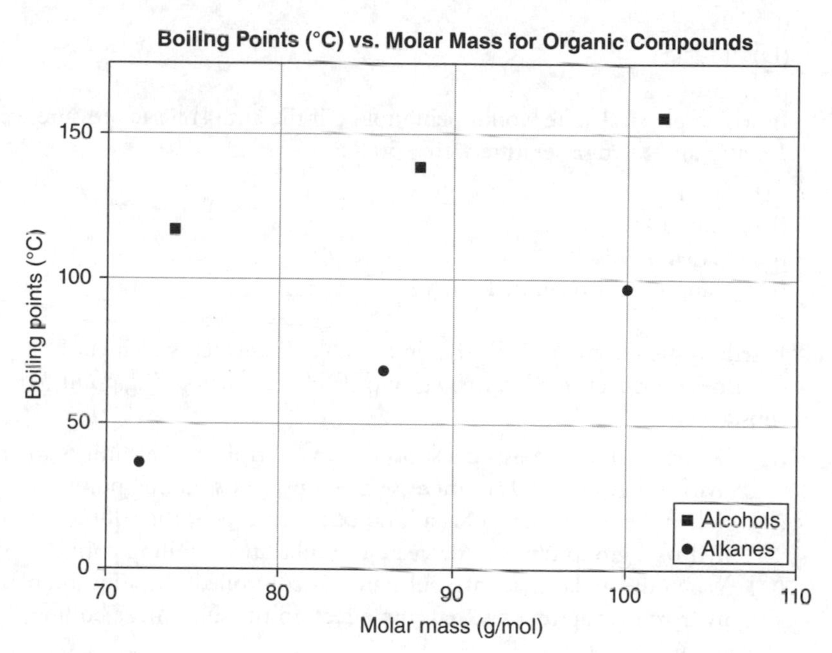

Figure 7.4

276. What is the boiling point of heptane?

(A) 36°C
(B) 98°C
(C) 117°C
(D) 156°C

277. Based on Figure 7.4, what is the best summary of the relationship between the molar mass and boiling point of an alkane?

(A) The molar mass of a substance seems to have little effect on its boiling point.
(B) As the molar mass of a substance increases, its boiling point tends to increase.
(C) As the molar mass of a substance increases, its boiling point tends to decrease.
(D) The relationship between the molar mass and boiling point of a substance is not clear from the data in the figure.

278. The organic substance propanol has a molar mass of about 60 g/mol. Which temperature is most likely the boiling point of propanol?

(A) 11°C
(B) 52°C
(C) 97°C
(D) 117°C

279. In which physical state would pentanol be if the atmospheric pressure were 1.0 atm and the temperature were 150°C?

(A) Solid
(B) Liquid
(C) Gaseous
(D) Cannot be determined

280. Based on the skeletal models shown in Table 7.2, which statement best describes the effect of a hydroxyl group (OH^-) on the boiling point of a substance?

(A) When the molar mass of a substance is controlled, the addition of a hydroxyl group tends to increase the substance's boiling point.
(B) When the molar mass of a substance is controlled, the addition of a hydroxyl group tends to decrease the substance's boiling point.
(C) When the molar mass of a substance is controlled, the addition of a hydroxyl group tends to have no effect on the substance's boiling point.
(D) The figure does not show a relationship between the presence of a hydroxyl group and the boiling point of a substance.

281. Based on the skeletal models shown in Table 7.2, which of the following images shows the skeletal model for hexanol?

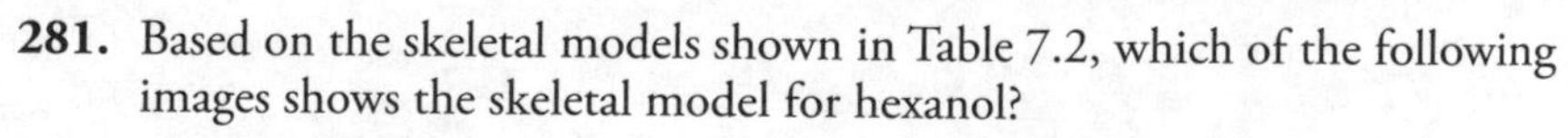

(A)
(B)

(C) OH
(D) OH

282. Organic chemists draw skeletal models to simplify the actual atoms and connections that exist within a molecule. Each short line segment represents the connection between two carbon atoms. In alkane molecules, hydrogen atoms surround each carbon atom in such a way that there are three hydrogen atoms on each end carbon and two hydrogen atoms on each middle carbon. These molecules can also be represented by ball-and-stick models or chemical formulas. An example for the molecule hexane is shown in Figure 7.5.

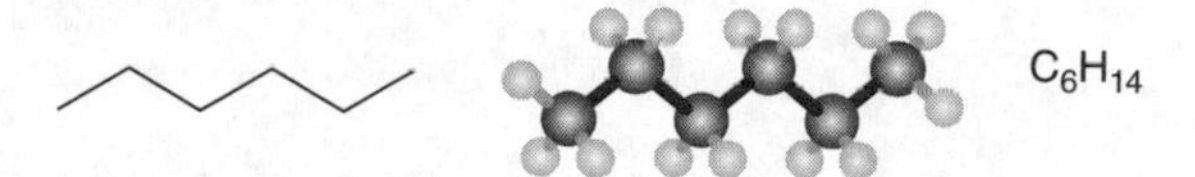

Figure 7.5

The alkane molecule decane has the following skeletal structure:

What is the chemical formula for decane?

(A) $C_{10}H_{20}$
(B) $C_{10}H_{22}$
(C) $C_{10}H_{24}$
(D) $C_{10}H_{30}$

283. Decane has a molar mass of 142 g/mol. What could a person expect its boiling point to be?

(A) 212°C
(B) 174°C
(C) 156°C
(D) 142°C

CHAPTER 8

Test 8

Passage 23

The sun is a source of many wavelengths of radiation that reach the earth. The earth's atmosphere absorbs some of these wavelengths, while others are able to penetrate and reach the planet's surface. Ultraviolet radiation from the sun comes in three different categories based on wavelength and penetration: UVA, UVB, and UVC.

UVB radiation has wavelengths of 280 to 320 nm and is partially absorbed by the earth's ozone layer. The UVB rays that do reach the surface can be absorbed by human skin and have been known to cause sunburn and many forms of skin cancer. Many products, from glasses to sunscreen, have been created to help protect humans from UVB radiation. Two groups of students set out to test the ability of materials to block UVB light, using a computer and a sensor specifically designed to detect UVB radiation.

Group 1

The members of Group 1 placed a sensor in full sunlight and shielded the sensor with a variety of sunglasses claiming to offer UVB protection. A reading was taken on the UVB sensor for each product, and the data were recorded in Table 8.1.

TABLE 8.1

	Approx. Retail Cost of Glasses ($)	UVB Reading Before Shielding (mW/m^2)	UVB Reading After Shielding (mW/m^2)	Shielding (%)
Pair A	10	742.3	2.7	99.6
Pair B	350	742.3	3.2	99.6
Pair C	1	742.3	4.2	99.4
Pair D	25	742.3	3.5	99.5
Pair E	90	742.3	2.1	99.7
Clear plastic	n/a	742.3	423.1	43.0

Group 2

The members of Group 2 placed a sensor in the sun and shielded that sensor with a piece of glass. They tested sunscreens of increasing SPF (sun protection factor) on the glass, and the data were recorded in Table 8.2.

TABLE 8.2

Description	SPF	UVB Reading with Sunscreen (mW/m²)	UVB Reading without Sunscreen (mW/m²)	Cost per Bottle of Sunscreen ($)
Dark tanning	4	72.1	742.3	8.99
Waterproof	8	35.6	742.3	4.99
Sport	15	20.2	742.3	7.99
Oil-free	30	19.8	742.3	15.99
Baby	50	18.5	742.3	10.99

284. In Group 2's experiment, SPF is the independent variable being manipulated and UVB is the dependent variable being measured. Which of the following graphs best represents the relationship between the SPF and UVB data from this experiment?

(A)

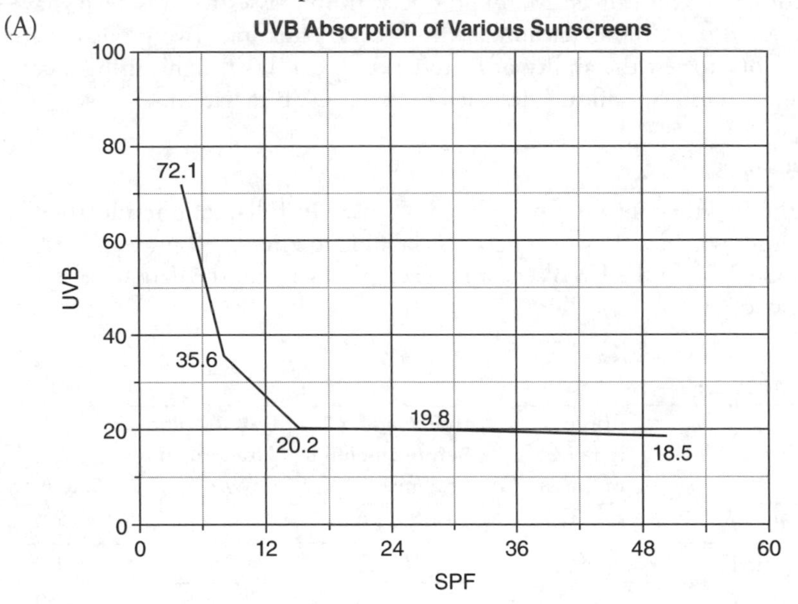

Figure 8.1

(B)

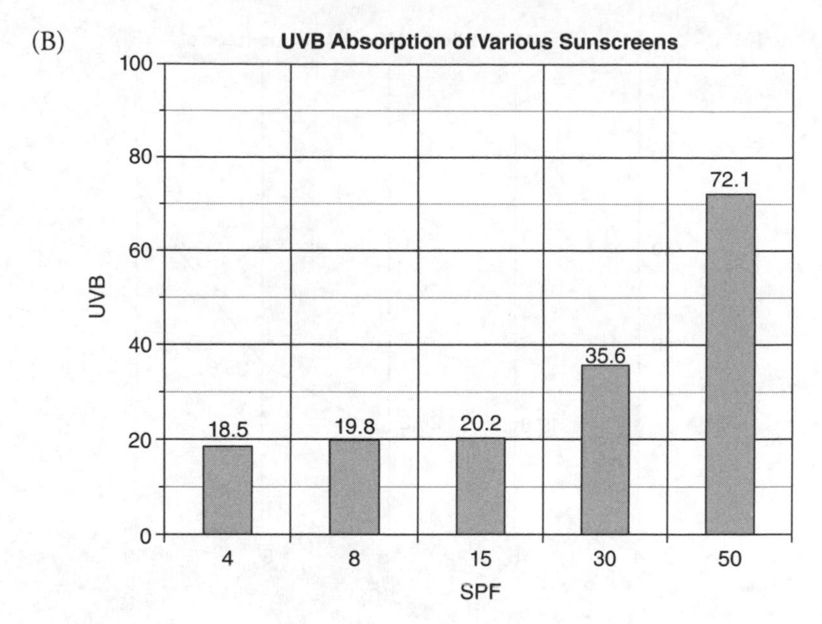

Figure 8.2

(C)

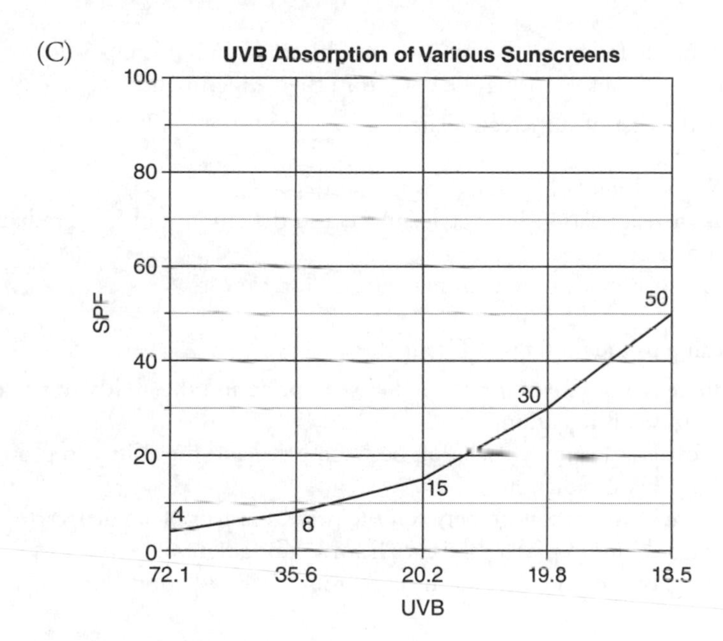

Figure 8.3

(D)

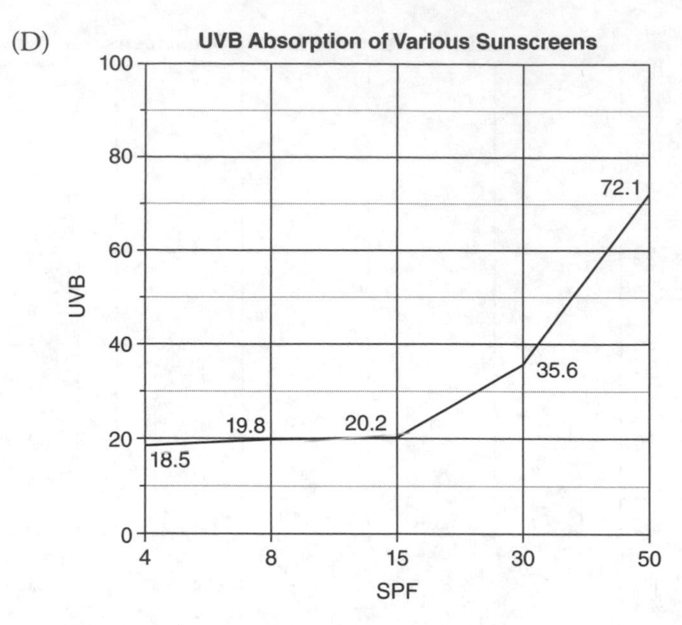

Figure 8.4

285. Which of the following questions could the students in Group 2 be attempting to answer using the data from their experiment?

(A) Is the cost of sunscreen related to the amount of UVB that is blocked?
(B) Which brand of sunscreen is the best?
(C) Is there a relationship between SPF and the amount of UVA radiation blocked?
(D) Can sunscreen protect humans from skin cancer?

286. According to the findings of Group 1:

(A) there is a negative correlation between price and the ability to protect from UVB radiation.
(B) there is a positive correlation between price and the ability to protect from UVB radiation.
(C) there is no correlation between the price of a pair of sunglasses and their ability to protect the eyes from UVB radiation.
(D) sunglasses do not protect the eyes from UVB radiation.

287. Assuming that the sunscreen being tested was purchased in 10 oz bottles, which sunscreen had the best cost for the amount of UVB protection (use the formula $/mW/m² of UVB blocked)?

(A) SPF 50
(B) SPF 30
(C) SPF 8
(D) SPF 4

288. Which of the following is MOST likely to represent a control that would have been used to ensure reliability of data in the experiment done by Group 2?

(A) The distance from the sunglasses to the sensor
(B) The ingredients in the sunscreen being used
(C) A longer time for the sensor readings as SPF increased
(D) The amount and thickness of sunscreen being spread on the glass

289. According to the results of Group 1's experiment, what percentage of UVB rays would a $200 pair of sunglasses block?

(A) 99.5%
(B) 99.7%
(C) 99.6%
(D) There is not enough information to determine this answer.

290. Based on the results of Group 2's experiment, what would the UVB reading most likely be if SPF 60 sunscreen were to be tested?

(A) 18 mW/m^2
(B) 19.25 mW/m^2
(C) 9.25 mW/m^2
(D) 4.5 mW/m^2

291. Which of the following is the experimental variable that Group 1 manipulated?

(A) The time of day
(B) The type of sunglasses
(C) The amount of UVB rays
(D) The distance of the materials from the sensor

292. Which of the following implies the correct relationship between SPF and UVB blockage?

(A) Sunscreens with SPFs higher than 30 provide only a marginal increase in sun protection over their counterparts with lower SPFs.
(B) As SPF increases, the ability to block UVB light decreases.
(C) There is no correlation between SPF and UVB blockage.
(D) Sunscreens with higher SPFs provide less sun protection than those with low SPFs.

293. What other experiment could the students in Group 2 conduct using the same equipment?

(A) The effect of sunscreen on shielding UVA rays
(B) The ability of sunscreen to prevent cancer
(C) How the amount of sunscreen applied can impact the UVB rating
(D) The margin of benefit of SPF sunscreens higher than 60

Passage 24

A study was conducted to identify the factors that affect the evaporation rates of various liquids in air. Throughout the experiment, the amount of liquid was varied, and the surface area exposed to the air was also manipulated. Table 8.3 displays the results.

The experiment was continued over a period of seven weeks for water and alcohol. Figure 8.5 shows this additional data graphically.

TABLE 8.3

Type of Liquid	Surface Area Exposed (cm^2)	Initial Amount (mL)	Amount Evaporated after 1 Week of Exposure (mL)			
			Trial 1	Trial 2	Trial 3	Average
Water	16	20.0	12.3	11.5	12.4	12.1
Water	12	40.0	9.3	9.2	8.8	9.1
Water	8	60.0	6.3	5.9	6.2	6.1
Water	4	80.0	2.8	3.1	2.9	2.9
Orange juice	4	80.0	2.7	3.2	3.2	3.0
Liquid bleach[1]	4	80.0	3.1	3.3	3.0	3.1
Vegetable oil	4	80.0	0.1	0.0	0.0	0.0
Rubbing alcohol	4	80.0	9.1	8.8	8.9	8.9
Rubbing alcohol	8	80.0	18.6	18.1	17.8	18.2
Rubbing alcohol	12	80.0	27.1	28.0	26.9	27.3
Rubbing alcohol	16	80.0	38.0	37.1	34.0	36.4

[1]The liquid bleach was approximately 5% sodium hypochlorite and 95% water.

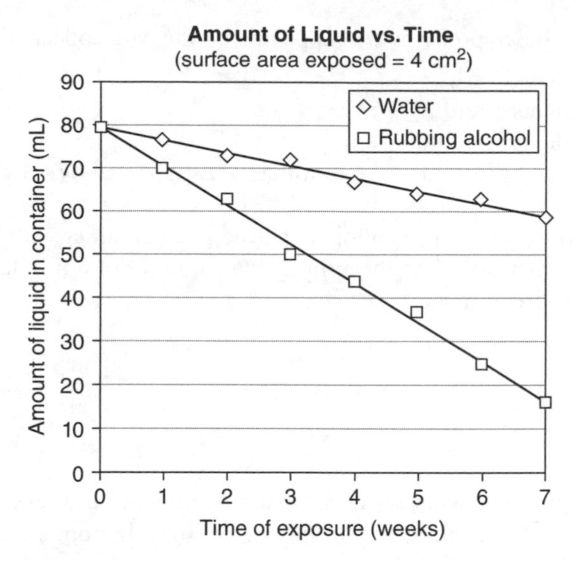

Figure 8.5

294. Which of the following statements do the data in Table 8.3 NOT support?

(A) The trials with larger surface areas of exposed water had greater evaporation rates.
(B) The evaporation rate for water is less than that for rubbing alcohol.
(C) Water had approximately the same evaporation rate as orange juice.
(D) Larger amounts of water correlate to higher evaporation rates.

295. In the experiment, 80 mL of orange juice with 4 cm^2 exposed was left in the open for one week. Using the data in Table 8.3, about how much of the original liquid was left at the end of the week?

(A) 3.0 mL
(B) 77.0 mL
(C) 80.0 mL
(D) 83.0 mL

296. Before the experiment, students made the following hypotheses:

Student 1: "Since I can smell rubbing alcohol as soon as I open the bottle, I expect it to have a greater rate of evaporation in air."

Student 2: "Since orange juice and liquid bleach are composed primarily of water, their evaporation rates will be close to that of water."

Student 3: "Surface area should not affect the rate of evaporation of a liquid because only the total amount of liquid affects evaporation rates."

Which of the hypotheses are supported by the data collected?

(A) Student 1 only
(B) Students 1 and 2
(C) Students 1, 2, and 3
(D) None of the students' hypotheses are supported by the data.

297. If 80.0 mL of rubbing alcohol are placed in a container with 20 cm^2 exposed, what would be the approximate amount of liquid left in the container after one week?

(A) 9 mL
(B) 15 mL
(C) 36 mL
(D) 45 mL

298. Using Figure 8.5, what is the approximate number of weeks required for 80 mL of rubbing alcohol to evaporate completely from a container with a 4 cm^2 exposure?

(A) 9 weeks
(B) 12 weeks
(C) 28 weeks
(D) The data do not provide enough evidence to make a reasonable prediction.

299. According to Figure 8.5, what is the rate of evaporation of water with a 4 cm^2 exposed surface area?

(A) 3 mL of water each week
(B) 10 mL of water each week
(C) 20 mL of water each week
(D) 80 mL of water each week

300. Which of the following conclusions may be supported by Figure 8.5?

(A) The rate of evaporation for alcohol increases with time.
(B) The rate of evaporation for alcohol decreases with time.
(C) The rate of evaporation for alcohol is fairly steady with time.
(D) The rate of evaporation for water is greater than that for alcohol.

301. If data for vegetable oil were added to Figure 8.5, one would most likely see:

(A) data with a steeper negative slope than that of rubbing alcohol.
(B) data very similar to the line for water.
(C) data with a flat line.
(D) data very similar to the line for rubbing alcohol.

302. Which of the following statements about rubbing alcohol is supported by the data?

(A) The variability between the trials increases with the surface area exposed.
(B) A decrease in surface area of exposure increases the evaporation rate.
(C) An increased amount of liquid in the container increases the evaporation rate.
(D) The rate of evaporation for rubbing alcohol is greater than that for ethyl alcohol.

303. Although the graph for rubbing alcohol displays a general downward trend, the variations in the data could possibly be attributed to all of the following EXCEPT:

(A) fluctuations in temperature in the room in which the containers were located.
(B) variations in the air flow in the room in which the containers were located.
(C) inaccuracies in the measurement of liquid volume.
(D) varying amounts of initial liquid in the containers.

Passage 25

A simple pendulum consists of a mass (the *pendulum bob*) suspended by a string, as shown in Figure 8.6. In an experiment, the mass of the bob, the radius of the arc, and the release height (measured vertically from the bottom of the swing) were varied. Rather than measuring the speed at the bottom of the swing, energy analysis was used to predict the speed of the pendulum bob at the bottom of the swing. The results are shown in Table 8.4.

A second experiment used the same scenario, but it included the measurement of the centripetal force and calculation of centripetal acceleration. *Centripetal force* is a real, unbalanced force pointed toward the center of an object's circular motion. Likewise, *centripetal acceleration* is defined as the component of acceleration directed toward the center. As a pendulum bob swings through

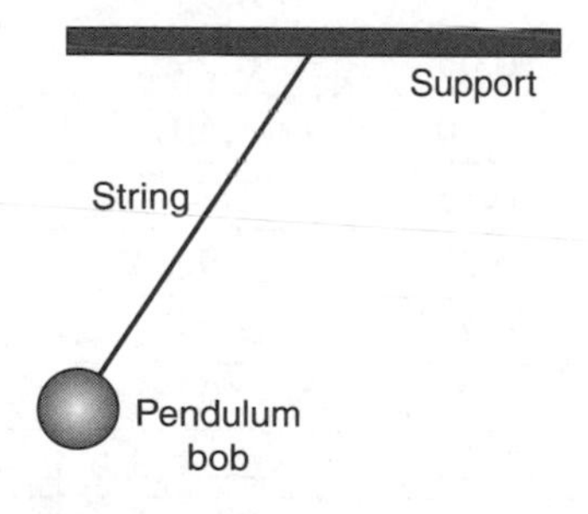

Figure 8.6

TABLE 8.4

Mass of Pendulum Bob (kg)	Radius of Arc (m)	Release Height (m)	Gravitational Energy at Top of Swing (J)	Kinetic Energy at Bottom of Swing (J)	Speed at Bottom of Swing (m/s)
0.010	0.05	0.05	0.005	0.005	0.99
0.010	0.10	0.05	0.005	0.005	0.99
0.010	0.20	0.05	0.005	0.005	0.99
0.010	0.30	0.05	0.005	0.005	0.99
0.010	0.40	0.05	0.005	0.005	0.99
0.010	0.40	0.10	0.010	0.010	1.40
0.010	0.40	0.15	0.015	0.015	1.71
0.010	0.40	0.20	0.020	0.020	1.98
0.010	0.40	0.25	0.025	0.025	2.21
0.020	0.40	0.25	0.049	0.049	2.21
0.030	0.40	0.25	0.074	0.074	2.21
0.040	0.40	0.25	0.098	0.098	2.21
0.050	0.40	0.25	0.123	0.123	2.21

the bottom of its arc, the string force dominates the gravitational force, thus providing the centripetal force that gives the pendulum bob its upward centripetal acceleration. The results are shown in Table 8.5.

TABLE 8.5

Mass of Pendulum Bob (kg)	Radius of Arc (m)	Release Height (m)	Speed at Bottom of Swing (m/s)	Centripetal Force (N)	Centripetal Acceleration (m/s/s)
0.010	0.05	0.05	0.99	0.196	19.6
0.010	0.10	0.05	0.99	0.098	9.8
0.010	0.20	0.05	0.99	0.049	4.9
0.010	0.30	0.05	0.99	0.033	3.3
0.010	0.40	0.05	0.99	0.025	2.5
0.010	0.40	0.10	1.40	0.049	4.9
0.010	0.40	0.15	1.71	0.074	7.4
0.010	0.40	0.20	1.98	0.098	9.8
0.010	0.40	0.25	2.21	0.123	12.3
0.020	0.40	0.25	2.21	0.245	12.3
0.030	0.40	0.25	2.21	0.368	12.3
0.040	0.40	0.25	2.21	0.490	12.3
0.050	0.40	0.25	2.21	0.613	12.3

304. According to the data in Table 8.4, increasing the mass of the pendulum bob:

(A) has no effect on the gravitational energy at the top of the swing.
(B) decreases the gravitational energy at the top of the swing.
(C) increases the radius of the arc.
(D) has no effect on the speed at the bottom of the swing.

305. A 0.010 kg pendulum has an arc radius of 0.40 m. Using the data trends shown in Table 8.4, predict the kinetic energy at the bottom of the swing if it is released from a height of 0.35 m.

(A) 0.025 J
(B) 0.030 J
(C) 0.035 J
(D) 0.040 J

306. According to Table 8.4, when the release height doubles, the gravitational energy at the top of the swing:

(A) doubles.
(B) quadruples.
(C) decreases to one-half its value.
(D) decreases to one-fourth its value.

307. Which of the following conclusions about energy is supported by Table 8.4?

(A) Kinetic energy at the bottom of the swing is directly proportional to speed.
(B) Gravitational energy at the top of the swing is inversely proportional to release height.
(C) Kinetic energy at the bottom of the swing is directly proportional to the radius of the arc.
(D) Gravitational energy at the top of the swing equals kinetic energy at the bottom of the swing.

308. When the mass of the pendulum bob doubles, the kinetic energy at the bottom of the swing:

(A) doubles.
(B) quadruples.
(C) decreases to one-half its value.
(D) decreases to one-fourth its value.

309. When the pendulum bob's kinetic energy doubles, its speed:
(A) doubles.
(B) decreases to one-half its value.
(C) increases by a factor of 1.4.
(D) increases by a factor of 2.2.

310. According to Table 8.5, centripetal acceleration is
(A) independent of mass.
(B) directly proportional to mass.
(C) inversely proportional to mass.
(D) directly proportional to the radius of the arc.

311. When the radius of the arc doubles, the centripetal force:
(A) doubles.
(B) quadruples.
(C) decreases to one-half its value.
(D) decreases to one-fourth its value.

312. A car approaches a school zone with a speed limit of 20 miles per hour. Using the data trends shown in Table 8.4, how does the kinetic energy of a car speeding at 40 miles per hour compare to that of a car moving at the speed limit?
(A) The speeding car's kinetic energy is one-half that of the other car.
(B) The speeding car's kinetic energy is one-fourth that of the other car.
(C) The speeding car's kinetic energy is twice that of the other car.
(D) The speeding car's kinetic energy is four times that of the other car.

313. Using Table 8.5, predict the centripetal force on a 0.060-kg bob with a 0.40-m arc radius that is released from a height of 0.25 m.
(A) 0.613 N
(B) 0.736 N
(C) 1.226 N
(D) 9.800 N

Passage 26

Phosphorus is an essential nutrient that can negatively affect water quality, primarily by promoting excessive plant and algae growth. When this occurs, plants and animals that live in the water are affected by the reduced sunlight and lower oxygen levels that develop as organic matter decomposes. For humans, algal blooms lead to a reduction in the quality of drinking water, a decrease in the use of the water source for recreational activities, and a decline in property value along waterfront areas.

Lakes are often classified according to their *trophic state*, which indicates their biological productivity. The least productive lakes are called *oligotrophic*. Bodies of water classified as oligotrophic are typically cool and clear, have relatively low nutrient concentrations, and provide excellent drinking water. The most productive lakes are called *eutrophic* and are characterized by high nutrient concentrations that result in algal growth, cloudy water, and low dissolved oxygen levels. Table 8.6 shows the phosphorus levels that are found in lakes with different trophic classifications.

TABLE 8.6

Trophic Classification	Phosphorus (µg/L)
Oligotrophic	0–12
Mesotrophic	12–24
Eutrophic	24–96
Hypereutrophic	96+

Lake managers collected data over approximately two decades in four different ecological areas of a large lake. Each area had a different target phosphorus level based on the natural ecological factors of the area, as indicated in Figures 8.7 through 8.10. For proper lake health, the level must be at or below the target amount. Levels in excess of the target amount lead to an imbalance in nutrient flow.

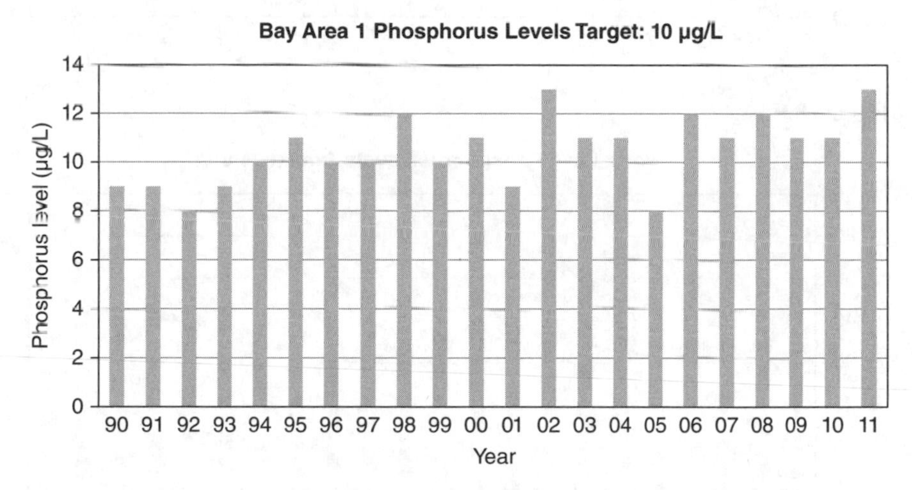

Figure 8.7

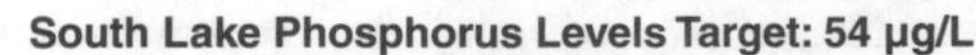

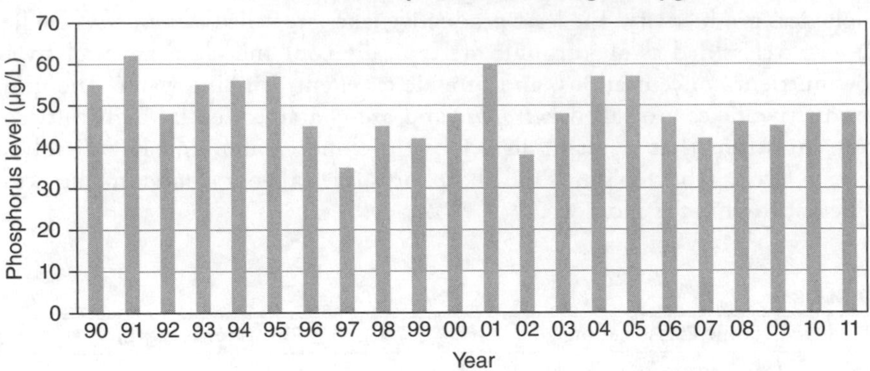

Figure 8.8

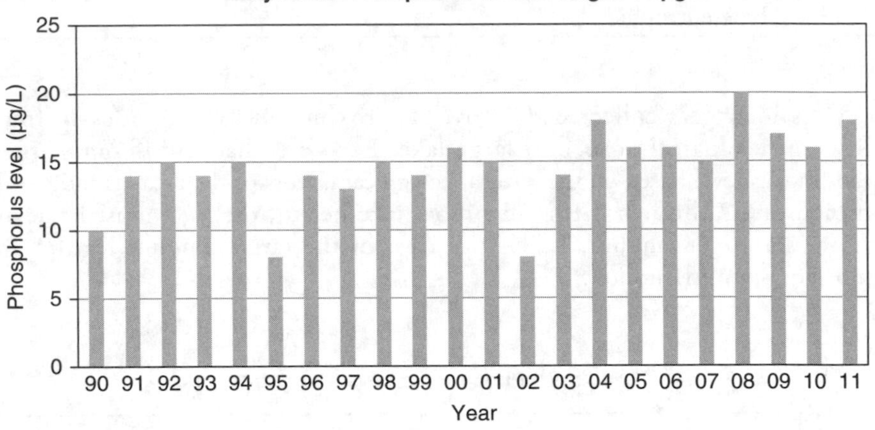

Figure 8.9

North Lake Phosphorus Levels Target: 25 µg/L

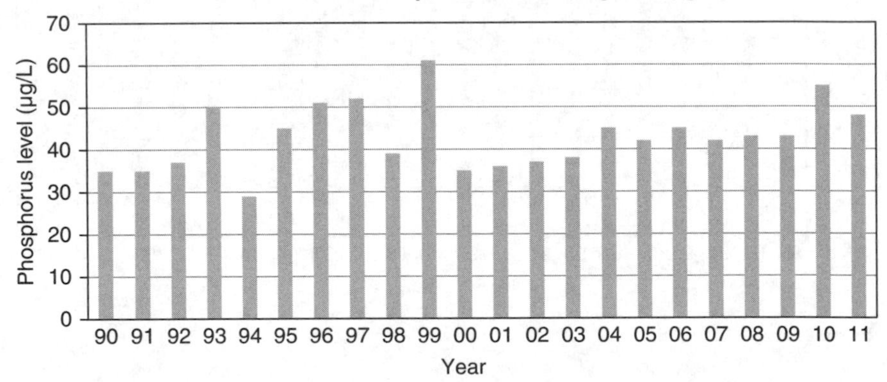

Figure 8.10

314. In which area of the lake did the scientists fail to attain the target phosphorus level in *any* of the years of the study?

(A) Bay Area 1
(B) North Lake
(C) Bay Area 2
(D) South Lake

315. The target for Bay Area 2 falls in the trophic category of:

(A) oligotrophic.
(B) mesotrophic.
(C) eutrophic.
(D) hypereutrophic.

316. In 2005, ecologists managing the lake began a concentrated effort to reduce agricultural runoff. This appears to have had the greatest effect on phosphorus levels in:

(A) Bay Area 1.
(B) North Lake.
(C) Bay Area 2.
(D) South Lake.

317. In 2002, the South Lake area had a trophic classification of:

(A) oligotrophic.
(B) mesotrophic.
(C) eutrophic.
(D) hypereutrophic.

318. For how many years of the study was Bay Area 2 found to be mesotrophic?

(A) 0
(B) 3
(C) 9
(D) 19

319. During which of the following years did Bay Area 1 have a trophic classification of mesotrophic?

(A) 1992
(B) 1995
(C) 2009
(D) 2011

320. Which of the following best describes the range in phosphorus levels in the North Lake over the 21-year period?

(A) 32 µg/L
(B) 62 µg/L
(C) 25 µg/L
(D) 45 µg/L

321. Which lake had the narrowest range of phosphorus levels over the time period of the study?

(A) Bay Area 1
(B) North Lake
(C) Bay Area 2
(D) South Lake

322. During how many years did Bay Area 1 meet or surpass the standard for proper lake health?

(A) 4
(B) 10
(C) 10
(D) 15

323. What was the difference in phosphorus concentration between Bay Area 1 and North Lake in 1993?

(A) 9 µg/L
(B) 41 µg/L
(C) 50 µg/L
(D) 59 µg/L

324. A neighboring lake was tested in 2005 and found to have a phosphorus level of 10 µg/L. It was most likely taken from a body of water similar to:

(A) Bay Area 1.
(B) North Lake.
(C) Bay Area 2.
(D) South Lake.

325. Which of the following statements is best supported by the information in the passage and figures?

(A) When studying the North Lake area, one would expect to find cool, clear water and high oxygen levels.
(B) None of the target levels for any of the lake areas fell in the eutrophic category.
(C) The management of phosphorus levels does not have a positive impact on humans who use the lake for recreation.
(D) During the 21-year period of the study, none of the lakes could be classified as hypereutrophic.

326. The category of lake classification appearing most frequently in the North and South Lake areas was:

(A) oligotrophic.
(B) mesotrophic.
(C) eutrophic.
(D) hypereutrophic.

327. According to the information provided in the passage, which of the areas was the most likely to have the lowest dissolved oxygen levels in 2010?

(A) Bay Area 1
(B) North Lake
(C) Bay Area 2
(D) South Lake

CHAPTER 9

Test 9

Passage 27

There is some evidence that ancient civilizations knew placing various metals together could create an electrical current. In the year 1800, Alexander Volta published experiments outlining his discovery of the *voltaic pile*, a device commonly referred to as the first electric battery. Volta stacked two different metals on either side of a wet felt disk and found that certain combinations produced an electrical voltage. By the early 1800s, many scientists were expanding on the idea of the voltaic pile by making apparatuses now known as *voltaic cells*. These cells generally contain two separate jars connected by a *salt bridge*, or porous membrane. Each jar contains a certain metal and a solution of the positive ions of the same metal. When different jars containing different metals are connected, an electrical voltage can be produced. A theoretical example of a copper/zinc voltaic cell is shown in Figure 9.1.

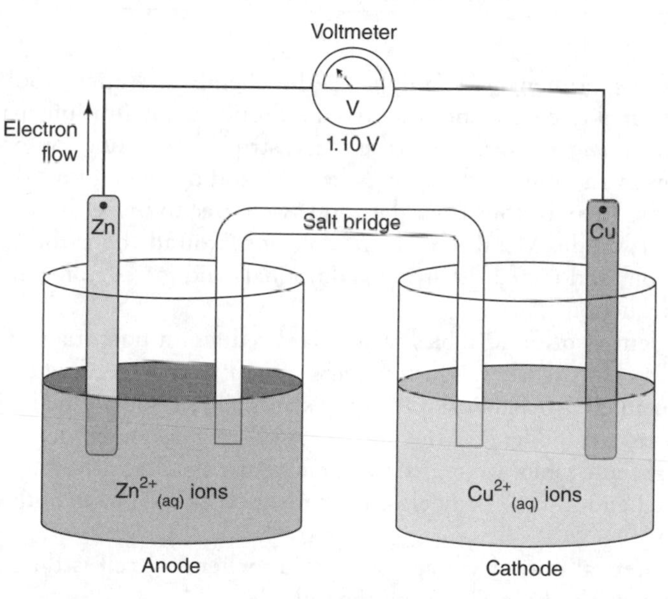

Figure 9.1

The last two centuries have seen a marked spike in demand for smaller batteries that can produce higher voltages for longer periods of time. This led a chemistry student to become interested in how using different metals in the voltaic cell could increase the voltage output of that cell. To conduct the experiment she chose four different types of metals that were available in strips from her local hardware store. These metals were zinc (Zn), lead (Pb), copper (Cu), and silver (Ag). She used 1-molar concentrated solutions of each of the various metal ions. She then made a salt bridge out of filter paper soaked in a potassium chloride brine solution. Many different combinations of metals were attempted, and the voltage output was measured with a standard voltmeter. The results of the experiment are recorded in Table 9.1.

TABLE 9.1

Experiment Number	Jar 1 Contents	Jar 2 Contents	Measured Output Voltage (V)
1	Zn metal/Zn^{2+}	Pb metal/Pb^{2+}	0.63
2	Zn metal/Zn^{2+}	Cu metal/Cu^{2+}	1.10
3	Zn metal/Zn^{2+}	Ag metal/Ag^{+}	1.56
4	Cu metal/Cu^{2+}	Ag metal/Ag^{+}	0.47
5	Cu metal/Cu^{2+}	Pb metal/Pb^{2+}	0.65
6	Pb metal/Pb^{2+}	Ag metal/Ag^{+}	0.73
7	Pb metal/Pb^{2+}	Pb metal/Pb^{2+}	0.00

After the experiment was completed, the chemistry student looked at the literature to make sense of her results. She found two definitions particularly helpful. The *anode* was defined as the metal strip where oxidation occurs. The metal atoms in the anode were losing electrons and dissolving into the solution as metal ions. The electrons from the anode were free to move through the wire toward the cathode. Metal ions in the solution around the cathode accepted those electrons and joined the strip as additional solid metal atoms, in a process known as reduction.

The student also found tables of standard reduction potentials (Table 9.2). These tables compared how much voltage should be produced when the metal is placed in an electrochemical cell with a standard electrode. One table showed each metal as a cathode, and the other showed each as an anode. The student learned that these tables were used to calculate the theoretical voltage output of any electrochemical cell. (Any electrochemical cell has to have both a cathode and an anode.)

The theoretical output voltage of an electrochemical cell is the sum of the standard potentials of the anode and the cathode.

TABLE 9.2

Cathode (Reduction)	Standard Potential (Volts)	Anode (Oxidation)	Standard Potential (Volts)
$Al^{3+}_{(aq)} + 3e^- \rightarrow Al_{(s)}$	−1.66	$Ag_{(s)} \rightarrow Ag^+_{(aq)} + e^-$	−0.80
$Zn^{2+}_{(aq)} + 2e^- \rightarrow Zn_{(s)}$	−0.76	$Cu_{(s)} \rightarrow Cu^{2+}_{(aq)} + 2e^-$	−0.34
$Pb^{2+}_{(aq)} + 2e^- \rightarrow Pb_{(s)}$	−0.13	$Pb_{(s)} \rightarrow Pb^{2+}_{(aq)} + 2e^-$	0.13
$Cu^{2+}_{(aq)} + 2e^- \rightarrow Cu_{(s)}$	0.34	$Zn_{(s)} \rightarrow Zn^{2+}_{(aq)} + 2e^-$	0.76
$Ag^+_{(aq)} + e^- \rightarrow Ag_{(s)}$	0.80	$Al_{(s)} \rightarrow Al^{3+}_{(aq)} + 3e^-$	1.66

328. Which of the following best describes the independent variables of this investigation?

(A) The types of metal used for the cathode and anode
(B) The type of metal ion solution used in the anode jar
(C) The amount of output voltage produced by various electrochemical cell configurations
(D) The concentration of metal ion solution used in each electrochemical cell

329. Which variable should not be controlled for this experiment?

(A) The surface area of the metal strip used for the anode
(B) The type of metal strip used for the anode
(C) The amount of time the electrochemical cell is allowed to operate
(D) The concentration of potassium chloride used to make the salt bridge

330. A lead/silver electrochemical cell is expected to have an output voltage of 0.93 V. However, Experiment 6 from Table 9.1 shows a measured output voltage of 0.73 V. Which of the following might account for the difference?

(A) Silver should have been the cathode, and lead should have been the anode, but the experimenter switched them.
(B) Experiment 6 was the third time the silver ion solution had been used, and the concentration of silver ions had been diminished.
(C) The experimenter mistakenly switched a lead strip with a zinc strip for the anode.
(D) The experimenter used a 1.5-inch-wide strip of lead as the anode instead of the standard 1-inch strip.

331. Experiment 1 from Table 9.1 shows a zinc/lead cell with a measured output voltage of 0.63 V. Which of the following best describes the cell?

(A) Zinc is the cathode and lead is the anode.
(B) Zinc is the cathode and zinc ions are the anode.
(C) Zinc is the anode and lead is the cathode.
(D) Zinc is the anode and zinc ions are the cathode.

332. Figure 9.1 shows the electrochemical cell the student used for Experiment 2 where zinc was the anode and copper was the cathode. What can be inferred about the mass of the zinc and copper strips as the experiment progressed?

(A) The mass of both strips increased.
(B) The mass of both strips decreased.
(C) The mass of the zinc strip increased, but the mass of the copper strip decreased.
(D) The mass of the zinc strip decreased, but the mass of the copper strip increased.

333. Experiment 8 was conducted with an aluminum strip in Jar 1 and a copper strip in Jar 2. Which metal would be the anode?

(A) Aluminum would be the anode.
(B) Copper would be the anode.
(C) Aluminum would be the anode until the metal dissolved to a certain point, and then copper would become the anode.
(D) There is not enough information to determine which would be the anode.

334. Experiment 8 was conducted with an aluminum strip in Jar 1 and a copper strip in Jar 2. What would be the expected output voltage of the electrochemical cell?

(A) 1.10 V
(B) 1.66 V
(C) 1.32 V
(D) 2.00 V

335. Further experimentation finds that when electrochemical cells are hooked together in a series configuration, their output voltage is added together for total output voltage. How many copper/zinc electrochemical cells are required to light a diode that needs a minimum of 12.0 V to operate?

(A) 2 cells
(B) 10 cells
(C) 11 cells
(D) 12 cells

Passage 28

Every time a lightbulb is switched on, an electrical circuit is formed. When plugged into an outlet that provides a certain voltage, current begins to flow through the bulb. Current depends on the resistance of the bulb and the voltage of the power supply. The bulb's power is a measure of the amount of energy the bulb requires each second. Power is calculated by multiplying current and voltage. Table 9.3 contains data that relate these variables for a basic circuit consisting of one lightbulb and a power supply.

TABLE 9.3

Resistance of Single Bulb (ohm)	Voltage across Bulb (V)	Current through Bulb (A)	Power Output (W)	Energy Usage/ Hour (kJ)
120	120	1.00	120	432
240	120	0.50	60	216
360	120	0.33	40	144
480	120	0.25	30	108
480	180	0.38	68	243
480	240	0.50	120	432
480	300	0.63	188	675

Figure 9.2 shows three configurations of bulbs. When the bulbs are connected in series, they form one path to the power supply. If any bulb in the pathway breaks, all the lights go out because the circuit is no longer complete. In contrast, bulbs connected in parallel are all independently connected to the power supply—in essence, forming their own circuits. Bulbs wired in parallel across a power supply continue to work even when one bulb goes out because each branch forms an independent circuit.

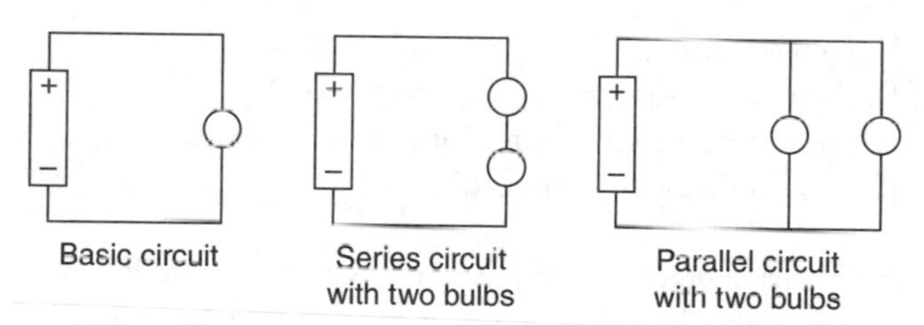

Figure 9.2

Table 9.4 shows how the number of bulbs in series and parallel affect the current and power values. It gives data for 240 ohm bulbs connected to a 120 V power supply.

TABLE 9.4

Arrangement of Bulbs	Number of Bulbs	Current in Each Bulb (A)	Current through Battery (A)	Power Output of Each Bulb (W)	Power Output of Circuit (W)
Series	2	0.25	0.25	15.0	30
Series	3	0.17	0.17	6.7	20
Series	4	0.13	0.13	3.8	15
Parallel	2	0.5	1.0	60	120
Parallel	3	0.5	1.5	60	180
Parallel	4	0.5	2.0	60	240

336. Based on the data in Table 9.3, describe the relationship between current, voltage, and resistance.

(A) Current is the ratio of voltage to resistance.
(B) Current is the product of voltage and resistance.
(C) Current is the ratio of resistance to voltage.
(D) Current is proportional to the square of voltage and independent of resistance.

337. Using information provided in the passage, determine the power output of a 50.0 ohm bulb connected to a 150 V socket that has 3.0 A flowing through it.

(A) 3 W
(B) 50 W
(C) 150 W
(D) 450 W

338. For a fixed voltage, what happens to the power output of a bulb when its resistance triples?

(A) The power triples.
(B) The power increases by a factor of 9.
(C) The power decreases to one-third of its value.
(D) The power decreases to one-ninth of its value.

339. For a bulb with a given resistance, what happens to the flow of current through the bulb when the voltage of the power supply doubles?

(A) The current doubles.
(B) The current quadruples.
(C) The current decreases to one-half its value.
(D) The current decreases to one-fourth its value.

340. A 480-ohm bulb is screwed into a 120-V socket. How much energy does it need to stay lit for four hours?

(A) 108 kJ
(B) 120 kJ
(C) 216 kJ
(D) 432 kJ

341. Predict the power required to operate a 600 ohm lightbulb when it is plugged into a 120 V outlet.

(A) 20 W
(B) 24 W
(C) 30 W
(D) 120 W

342. According to Tables 9.3 and 9.4, when a fifth bulb is added to a parallel circuit, each bulb will:

(A) output the same amount of energy per second as a bulb in a basic circuit.
(B) output more energy per second than a bulb in a basic circuit.
(C) output less energy per second than a bulb in a basic circuit.
(D) make the other bulbs get brighter.

343. When a fifth bulb is added to a series circuit, how will the bulb's power output compare to that of a bulb in a four-bulb series circuit?

(A) The fifth bulb produces the same amount of power.
(B) The fifth bulb produces twice the power.
(C) The fifth bulb produces less power.
(D) The fifth bulb produces no power.

344. According to the information in Tables 9.3 and 9.4, adding an additional bulb to a parallel circuit:

(A) increases the circuit's power output by decreasing the total resistance of the entire circuit.
(B) decreases the circuit's power output by decreasing the total resistance of the entire circuit.
(C) increases the circuit's power output by increasing the total resistance of the entire circuit.
(D) decreases the circuit's power output by increasing the total resistance of the entire circuit.

345. The circuit in Figure 9.3 shows Bulbs 2 and 3 wired in parallel. That combination is wired in series with Bulb 1 and the battery. Which of the following statements is FALSE?

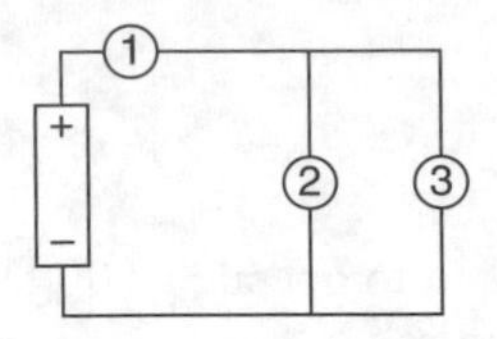

Figure 9.3

(A) If Bulb 1 breaks, the other two bulbs will go out.
(B) If Bulb 2 breaks, the other two bulbs will go out.
(C) If both Bulbs 2 and 3 break, Bulb 1 will go out.
(D) If Bulb 3 breaks, the other bulbs will stay lit.

346. As bulbs are added to a parallel circuit:

(A) more current flows through each bulb.
(B) less current flows through each bulb.
(C) less power is output from the circuit.
(D) more current flows through the battery.

347. A child noticed that five bulbs in her electric toy went out simultaneously, but four other bulbs remained lit. What is the most likely circuit arrangement in the toy?

(A) The five bulbs that went out are wired in parallel.
(B) Each of the five bulbs that went out are broken.
(C) The five bulbs that went out are wired in series.
(D) All nine bulbs are wired in parallel.

348. Adding additional bulbs to a series circuit:

(A) increases the resistance of the entire circuit.
(B) decreases the resistance of the entire circuit.
(C) decreases the power output of the entire circuit.
(D) increases the flow of current in each bulb.

349. A watt (W) of power is the total joules (J) of electrical energy transferred by a circuit element each second. Given a circuit with three 240 ohm bulbs wired in series to a 120 V power supply, how much energy is transferred by the circuit if it operates for 10 seconds?

(A) 6.7 J
(B) 67 J
(C) 150 J
(D) 200 J

350. How much power does a parallel circuit require if it has six 240-ohm bulbs connected to a 120 V power supply?

(A) 60 W
(B) 120 W
(C) 360 W
(D) 1,440 W

Passage 29

Since ancient times, scientists, philosophers, and other thinkers considered the smallest pieces of matter to be tiny spherical structures called atoms that were stacked in various arrangements. The ancient Greek philosopher Democritus also suggested that these atoms were indivisible and indestructible. The turn of the twentieth century brought with it a renewed interest in exploring atoms. Two groundbreaking experiments challenged the idea that atoms are the smallest constituent of matter. These experiments were very different, but both pointed to the fact that atoms contained even smaller parts.

Experiment 1: The 1897 Cathode Ray Tube Experiments

Scientists in the late 1890s investigated a curious new device known as a cathode ray tube (CRT). The CRT consisted of a glass tube with a metal wire coming out of each end. When all of the air was removed from the tube and a voltage was applied across the wires, mysterious green rays appeared at one end of the tube. These rays were called cathode rays. It was not immediately known if these rays were a type of wave or a type of particle, but several experiments were conducted on them. First it was discovered that the rays would always be attracted to an area of excess positive charge and away from an area of excess negative charge. Precise measurements could not determine the exact mass of the cathode rays, but it was determined that they did have mass. After that discovery, the rays were considered a particle instead of a wave.

The mass-to-charge ratio of the cathode rays was determined to be more than 1,000 times smaller than the same ratio for any known atom or ion. After subsequent experiments, researchers determined that cathode rays were small particles that broke off of an atom when a voltage was applied.

Experiment 2: The 1911 Gold Foil Experiment

At the time, the accepted model of the atom was a relatively solid sphere, similar to a ball of chocolate chip cookie dough. An experiment in 1911 brought that model into question. A beam of alpha particles (small, high-energy, positively charged particles) was shot into a piece of gold foil approximately 8.6×10^{-8} m in thickness. The alpha particles were thought to have enough energy to pass straight through the foil and hit a detector on the other side, and most of the particles did just that. However, a small fraction of alpha particles were deflected a few degrees as they passed through the foil. Upon closer examination of the data, a more startling fact was found—some alpha particles never hit the detector.

More detectors were added around the gold foil, and it was discovered that a tiny portion of the alpha particles, 1 out of every 20,000 particles, was deflected 90 degrees or more from the beam. Some particles even bounced straight back toward the alpha particle source. The scientist was so surprised by the results that he stated, "It was as if you fired a 15-inch shell at a sheet of tissue paper and it came back to you." It was concluded that there must be some particle inside an atom causing these major deflections of alpha particles.

351. Which of the following best describes what each experiment concluded about the newly discovered small particles that make up atoms?

	Experiment 1	Experiment 2
(A)	Dense and sturdy	Positively charged
(B)	Negatively charged	Dense
(C)	Positively charged	Negatively charged
(D)	Small	Low energy

352. Which of the following is the best conclusion concerning why the cathode rays in Experiment 1 had a mass-to-charge ratio 1,000 times smaller than that of any known atom?

(A) Cathode ray particles are 1,000 times less massive than any known atom.
(B) Cathode ray particles have 1,000 times less charge than any known atom.
(C) Cathode ray particles have 1,000 times more charge than any known atom.
(D) Cathode ray particles could either be less massive or have a greater charge than any known atom.

353. Modern chemistry books discuss several subatomic particles. Figure 9.4 offers a common diagram of such particles. The symbol (–) means negative, (+) means positive, and (Ø) means neutral. Which particle would have an undefined mass-to-charge ratio?

(A) Electron
(B) Proton
(C) Neutron
(D) Nucleus

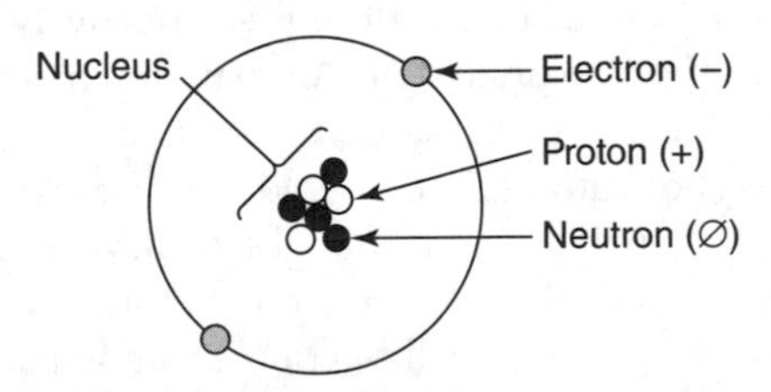

Figure 9.4

354. Which of the following statements do the data from both experiments support?

(A) Atoms are highly charged particles.
(B) Atoms are composed of a dense core known as a nucleus.
(C) Atoms are not indestructible and indivisible.
(D) The mass-to-charge ratio of atoms is much smaller than was originally thought.

355. Atoms, once thought to be solid spheres, have instead been proven to be mostly empty space with just a few tiny particles giving each atom its properties. Which data from either Experiment 1 or Experiment 2 best explains this fact?

(A) Experiment 1: Cathode rays are attracted to areas of positive electric charge.
(B) Experiment 1: Cathode rays have a small mass-to-charge ratio.
(C) Experiment 2: Some alpha particles were deflected from the beam at wide angles.
(D) Experiment 2: One out of every 20,000 particles got deflected to a large extent.

356. A micrometer (μm) is a common unit to measure small objects. One micrometer is equivalent to 0.000001 m. How thick was the gold foil in Experiment 2?

(A) 8,600 μm
(B) 86 μm
(C) 0.86 μm
(D) 0.086 μm

357. Which scientists would be most likely to write the following in their lab notebook: "I can see no escape from the conclusion that they are charges of negative electricity carried by particles of matter"?

(A) Scientists working on Experiment 1
(B) Scientists working on Experiment 2
(C) Scientists working on either experiment
(D) Scientists working on both experiments

358. Which headline best matches the conclusions of Experiment 2?

(A) Gold Foil: Alpha Particles Are Found Inside Atoms
(B) Cathode Rays Destroy Atoms
(C) Gold Foil: Solid Particle in Atom Incredibly Small
(D) Cathode Rays: Most Go Through But a Few Bounce Off

Passage 30

The moon is the earth's only natural satellite and is the fifth-largest moon in the solar system. Believed to be around 4.5 billion years old, the moon was a deity worshipped by ancient civilizations and the first object in the solar system besides Earth that human beings set foot on. It has played a special role in human history, yet despite its ubiquity, the origin of this celestial body remains surrounded by mystery. Theories about the origin of the moon have long been debated among scientists. The following are summaries of the most recent major theories of moon formation. The first three theories are known as the Big Three and represent the predominant ideas before the mid-1970s. The fourth theory represents the most recent school of thought, developed in light of evidence gained from the Apollo space program.

Theory 1: Fission

The moon was spun off from the earth when the planet was young and rotating rapidly on its axis. The empty space the moon left behind became the Pacific Ocean basin.

Theory 2: Capture

The moon formed elsewhere in the universe. At some point, it came near enough to be affected by the earth's gravitational field. The moon was pulled into permanent orbit by the earth's gravity.

Theory 3: Coaccretion

The earth and moon and all other bodies of the solar system condensed independently out of the huge cloud of cold gases and solid particles that constituted the primordial solar nebula. The moon then fell into orbit around the earth.

Theory 4: Giant Impact

The earth was struck by a body about the size of Mars very early in its history. A ring of debris from the impact containing primarily Earth materials and some materials from the impacting object eventually coalesced to form the moon.

359. In which of the theories would the rocks on the moon NOT necessarily bear any similarity to the rocks on the earth?

(A) Fission
(B) Capture
(C) Both fission and coaccretion
(D) Both capture and fission

360. In which *two* of the theories would the rocks on the moon be nearly identical to those on the earth?

(A) Capture and fission
(B) Capture and coaccretion
(C) Coaccretion and fission
(D) Giant impact and capture

361. Which of the following statements would best support the argument of a proponent of the capture theory?

(A) Planets are incredibly small compared to the vastness of space.
(B) Jupiter and Saturn (the giant gas planets) have captured moons.
(C) The moon and the earth have the same oxygen isotope composition.
(D) The moon does not have a regular-size core.

362. The Big Three theories have cleverly been dubbed the Daughter theory, the Sister theory, and the Spouse theory by scientists who compared the relationship of the moon and the earth to familial relationships. Based on the information in the passage, which of the following would most accurately associate the theories to their nickname: Daughter, Sister, and Spouse?

(A) Fission, coaccretion, and capture
(B) Capture, fission, and coaccretion
(C) Capture, coaccretion, and fission
(D) Fission, capture, and coaccretion

363. The following description of the formation of the moon is from a children's radio program in the 1930s. Which of the four moon theories does this seem to illustrate?

Once upon a time—a billion or so years ago—when the earth was still young—a remarkable romance developed between the earth and the sun—according to some of our ablest scientists. . . . In those days, the earth was a spirited maiden who danced about the princely sun, was charmed by him, yielded to his attraction, and became his bride. . . . The sun's attraction raised great tides upon the earth's surface . . . the huge crest of a bulge broke away with such momentum that it could not return to the body of Mother Earth. And this is the way the moon was born!

(A) Fission
(B) Capture
(C) Coaccretion
(D) Giant impact

364. The fission theory is refuted by which of the following pieces of evidence?
(A) The moon lacks a large core.
(B) There is a striking similarity between the oxygen isotopes present on the earth and those on the moon.
(C) Studies of isotopes found in rocks put the age of the earth and moon at 4.5 billion years.
(D) The Pacific Ocean basin was formed 70 million years ago.

365. The moon's crust is thinner on the side nearest the earth. Scientists believe that this is because the moon was close to the earth when it formed. As the moon's mantle cooled, the earth's gravitational field pulled slightly more mantle closer to the planet before it "set." A thicker mantle made for a thinner crust on the side nearest the earth. This piece of evidence contradicts the capture theory because:
(A) in the capture theory, the moon broke off from the earth.
(B) in the capture theory, the moon formed close to the earth.
(C) in the capture theory, the moon formed in another part of the solar system.
(D) this piece of evidence supports the coaccretion theory.

366. The evidence that rock samples from the moon match rocks from the earth's crust and mantle but some samples bear no resemblance to the earth rock best supports which of the following theories?
(A) Fission
(B) Capture
(C) Coaccretion
(D) Giant impact

367. The giant impact theory:
(A) is not likely to change with the discovery of new evidence.
(B) completely explains the origin of the moon.
(C) is unable to account for why the moon is made mostly of rock.
(D) is the theory best supported by the most current scientific evidence.

368. Which of the theories is best supported by the evidence that the earth and moon are both 4.5 billion years old and provide isotopic evidence that indicates they were formed in the same "neighborhood" of the solar system?
(A) Fission
(B) Capture
(C) Coaccretion
(D) Giant impact

369. A fifth theory of moon formation, called the colliding planetesimals theory, exists. In this theory, an asteroid-like chunk of rock orbiting the sun collided with an asteroid-like chunk of rock orbiting the earth. The moon then condensed from the debris of this collision. This theory would be weakened by which of the following pieces of evidence?

(A) Moon rock matches the rock from the earth's crust and mantle.
(B) The moon's crust is thinner on the side nearest the earth.
(C) The moon is made mostly of rock.
(D) Isotopes indicate that the earth and moon formed in the same area of the solar system.

CHAPTER 10

Test 10

Passage 31

When the effect of air on a falling object is negligible and gravity is the only significant force on that object, the object is considered to be in free fall. A scientist can create a free-fall scenario by removing all air from a chamber (thus creating a vacuum) and allowing an object to drop freely. When objects are not falling freely, air affects them in different ways, depending on variables such as speed, mass, and size. If objects are able to fall for enough time through the air, they will eventually reach *terminal velocity*, a point at which their velocity stops increasing. Table 10.1 and Figures 10.1 and 10.2 show the effect of air on falling balls of different mass and radius.

TABLE 10.1

	10 g Ball Freefalling (R = 1 cm)			10 g Ball Falling with Air (R = 1 cm)		
Time (s)	Distance Fallen (m)	Velocity (m/s)	Air Drag Force (N)	Distance Fallen (m)	Velocity (m/s)	Air Drag Force (N)
0	0.0	0.0	0	0.0	0.0	0.000
1	4.9	9.8	0	4.8	9.5	0.009
2	19.6	19.6	0	19.1	18.7	0.033
3	44.1	29.4	0	41.0	25.1	0.061
4	78.4	39.2	0	68.1	28.9	0.080
5	123.0	49.0	0	97.8	30.7	0.090
6	176.0	58.8	0	129.0	31.5	0.095
7	240.0	68.6	0	161.0	31.8	0.097
8	314.0	78.4	0	192.0	31.9	0.098
9	397.0	88.2	0	224.0	32.0	0.098
10	490.0	98.0	0	256.0	32.0	0.098
11	593.0	108.0	0	288.0	32.0	0.098
12	706.0	118.0	0	320.0	32.0	0.098

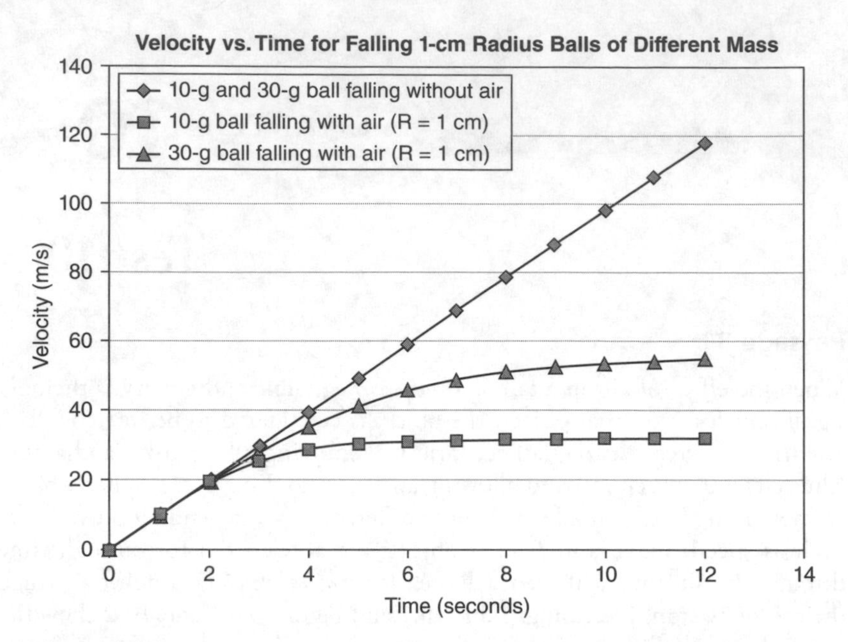

Figure 10.1

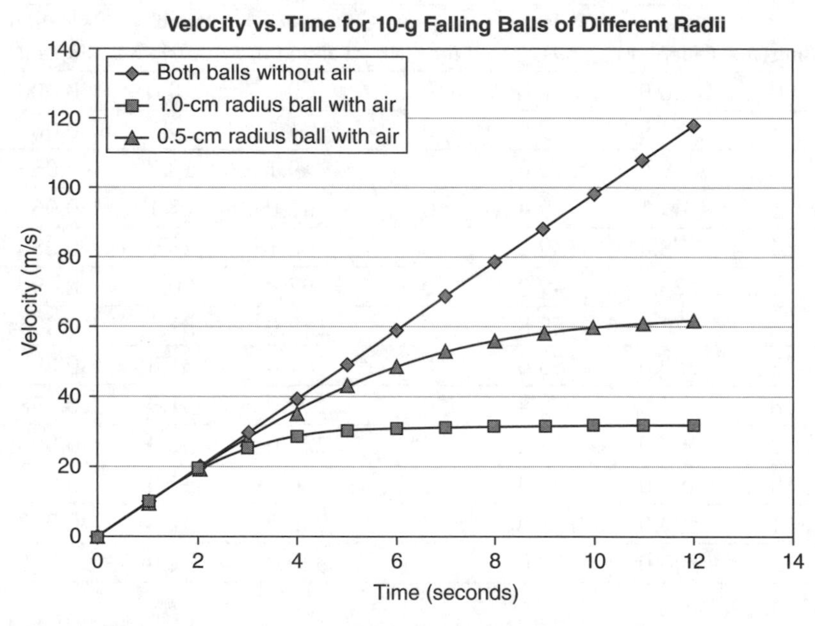

Figure 10.2

370. According to Table 10.1, which of the following is a correct statement about the velocity of a 10 g free-falling ball?

(A) The velocity is constant.
(B) The velocity increases 9.8 m/s each second.
(C) The velocity increases 4.9 m/s each second.
(D) The velocity steadily decreases.

371. According to Table 10.1, which of the following is a correct statement about the distance a 10 g ball free-falls?

(A) The distance increases constantly.
(B) The distance increases 9.8 m each second.
(C) The distance increases more each second compared to the previous second.
(D) The distance increases 4.9 m each second.

372. According to Table 10.1, how much does the air drag force on the ball (10 g, 1 cm) increase during the fourth second of falling through air?

(A) 0.000 N
(B) 0.019 N
(C) 0.061 N
(D) 0.080 N

373. Using the trends in Table 10.1, predict the velocity of the 10 g ball at the instant it has fallen freely from rest for 14 seconds.

(A) 9.8 m/s
(B) 19.6 m/s
(C) 118.0 m/s
(D) 138.0 m/s

374. After comparing the data for the falling ball with and without air, which of the following is a correct statement about the effect of air on velocity?

(A) The presence of air reduces the rate at which velocity increases.
(B) The presence of air increases the rate at which velocity increases.
(C) When air is not present, the velocity eventually stops increasing.
(D) When air is present, the velocity eventually stops decreasing.

375. For the 10 g ball falling with air, Table 10.1 indicates that the force of air drag:

(A) increases steadily with time.
(B) plateaus initially and then decreases.
(C) increases more each second compared to the previous second.
(D) increases more rapidly early in the fall and eventually plateaus.

376. According to the graph in Figure 10.1, what is the effect of mass on the velocity of a falling object in air from t = 2 to t = 6 seconds?

(A) Smaller masses lose speed more rapidly than larger masses.
(B) Larger masses lose speed more rapidly than smaller masses.
(C) Smaller masses gain speed more rapidly than larger masses.
(D) Larger masses gain speed more rapidly than smaller masses.

377. As an object falls, the effect of air on its motion:

(A) is more pronounced during the first few seconds.
(B) is more pronounced later in the fall.
(C) is minimal.
(D) eventually dissipates.

378. According to Figures 10.1 and 10.2, the terminal velocity of the 10 g ball with the 0.05 cm radius is most nearly:

(A) 9.8 m/s.
(B) 32 m/s.
(C) 55 m/s.
(D) 62 m/s.

379. In the absence of air drag, a ball with a larger radius:

(A) has the same motion as one with a smaller radius.
(B) reaches terminal velocity sooner than one with a smaller radius.
(C) obtains a larger terminal velocity.
(D) obtains a smaller terminal velocity.

380. The ball with less mass (but the same radius) falling through the air:

(A) reaches terminal velocity later than a ball with more mass.
(B) reaches terminal velocity within the first second it is dropped.
(C) reaches terminal velocity sooner than a ball with more mass.
(D) has the same terminal velocity as the ball with more mass.

381. Suppose one skydiver jumps out of an airplane feet first toward the ground. Her identical twin sister jumps out at the same time with her arms and legs extended and her stomach parallel to the ground. According to the passage, what can you infer about their subsequent motion?

(A) The feet-first twin will reach terminal velocity first.
(B) The belly-first twin will reach terminal velocity first.
(C) Both twins will reach terminal velocity at the same time.
(D) The belly-first twin will obtain a greater terminal velocity.

382. A table-tennis ball with a mass of 3 g is dropped side by side with an identical-sized wooden ball with a mass of 30 g. Based on the information provided, what can you conclude if the balls fall in a vacuum (no air)?

(A) The 30 g ball will hit the ground first.
(B) The 3 g ball will hit the ground first.
(C) Both balls will hit the ground at the same time.
(D) The 3 g ball will speed up more at first.

383. A table-tennis ball with a mass of 3 g is dropped side by side with an identical-sized wooden ball with a mass of 30 g. If both balls are dropped from rest and fall through the air, what can you conclude?

(A) The 30 g ball will hit the ground first.
(B) The 3 g ball will hit the ground first.
(C) Both balls will hit the ground at the same time.
(D) The 3 g ball will speed up more at first.

384. According to the data trends illustrated in the passage, a ball dropped in a vacuum without air will:

(A) eventually reach terminal velocity as it's falling.
(B) speed up at a steady rate as long as it has room to drop.
(C) speed up at greater rate if it's more massive.
(D) speed up at lesser rate if it's more massive.

Passage 32

A chemistry student was given the task of determining what percentage of a hydrated magnesium sulfate sample was water. Many chemical compounds exist in nature as hydrates instead of in a dry (or anhydrous) state. A sample of hydrated magnesium sulfate ($MgSO_4 \cdot nH_2O$) is a solid white powder that contains a certain number (n) of water molecules bonded to each magnesium sulfate crystal. One accepted method of isolating magnesium sulfate from the water molecules is by heating the sample in an open porcelain crucible. Magnesium sulfate has a very high boiling temperature, but water molecules can easily be turned into gas. When sufficiently heated, the water molecules are driven away from the container, leaving only anhydrous magnesium sulfate ($MgSO_4$).

The student cautiously heated the sample over a Bunsen burner. He then removed the crucible, allowed it to cool, and measured the mass. He returned the crucible to the burner and heated it again, allowed it to cool, and measured the mass again. The student repeated this procedure until he was certain all of the water had been removed. His data are found in Table 10.2.

TABLE 10.2

Mass of empty crucible	26.449 g
Mass of crucible + $MgSO_4 \cdot nH_2O$ sample	32.569 g
Mass of crucible and sample after first heating	32.558 g
Mass of crucible and sample after second heating	31.943 g
Mass of crucible and sample after third heating	29.439 g
Mass of crucible and sample after fourth heating	29.440 g

385. What mass of hydrated magnesium sulfate sample did the student place in the crucible at the start of the experiment?

(A) 32.569 g of $MgSO_4 \cdot nH_2O$
(B) 6.120 g of $MgSO_4 \cdot nH_2O$
(C) 29.440 g of $MgSO_4 \cdot nH_2O$
(D) 3.129 g of $MgSO_4 \cdot nH_2O$

386. How was the student certain that all of the water had been removed from the sample after the fourth heating?

(A) He had heated the sample for a sufficient amount of time.
(B) The mass of the sample and empty crucible remained at less than 30.00 g for two consecutive heatings.
(C) The mass did not change appreciably between the third and fourth heatings.
(D) The sample would have started to gain mass if more heating had occurred.

387. Which equation is the appropriate way to determine the percent mass of water in a sample of hydrated magnesium sulfate?

(A) $\dfrac{\text{Mass of water driven off}}{\text{Mass of sample before heating}} \times 100$

(B) $\dfrac{\text{Mass of water driven off}}{\text{Mass of anhydrous } MgSO_4} \times 100$

(C) $\dfrac{\text{Mass of water driven off}}{\text{Mass of crucible and sample after fourth heating}} \times 100$

(D) $\dfrac{\text{Mass of sample before heating}}{\text{Mass of water driven off}} \times 100$

388. From the data in Table 10.2, which is closest to the percent mass of the sample that is made up of water?

(A) 10% water molecules
(B) 25% water molecules
(C) 50% water molecules
(D) 90% water molecules

389. How would the calculated percent mass of water in the sample be affected if some of the sample splattered out of the crucible while it was being heated?

(A) The calculated percent mass of water would be too high.
(B) The calculated percent mass of water would be too low.
(C) The calculated percent mass of water would be too high or too low, depending on outside conditions.
(D) The calculated percent mass of water would remain unchanged.

390. The student left the crucible and the sample on the balance after the fourth heating. Forty minutes later, he came back and noticed the mass of the sample had changed again, as shown in Table 10.3. What is the best conclusion the student can reach about the additional change in mass?

(A) More water molecules were able to evaporate during the additional 40 minutes.
(B) During the additional 40 minutes, the contents cooled to room temperature, which caused the mass to increase.
(C) Water molecules were able to reattach to the $MgSO_4$ crystals during the additional 40 minutes.
(D) The magnesium sulfate crystals had time to expand during the additional 40 minutes.

TABLE 10.3

Mass of empty crucible	26.449 g
Mass of crucible + $MgSO_4 \cdot nH_2O$ sample	32.569 g
Mass of crucible and sample after first heating	32.558 g
Mass of crucible and sample after second heating	31.943 g
Mass of crucible and sample after third heating	29.439 g
Mass of crucible and sample after fourth heating	29.440 g
Mass of crucible and sample 40 minutes later	32.139 g

391. The student repeated the procedure but placed a lid on the crucible during the entire experiment. Which statement best describes the expected results?

(A) More water would be driven off because the crucible would get to a higher temperature.
(B) Drops of water would be found on the lid, but the mass of the sample would not change.
(C) Less water would be driven off because the heat could not get to the sample.
(D) The mass change of this experiment would be identical to the change in mass recorded in the initial experiment.

392. The student did a follow-up experiment by completely dissolving 50 g of anhydrous $MgSO_4$ in 100 mL of distilled water in Beaker 1 and 50 g of $MgSO_4 \cdot 7H_2O$ hydrate in 100 mL of distilled water in Beaker 2. How would the concentration of magnesium (Mg) in each beaker compare?

(A) Both beakers would have identical concentrations of magnesium.
(B) Beaker 1 would have a higher concentration of magnesium.
(C) Beaker 2 would have a slightly higher concentration of magnesium.
(D) Beaker 2 would have a concentration approximately 7 times higher than that of magnesium.

Passage 33

Astronomers have identified more than 170 moons in the solar system. For centuries, many of these scientists have used telescopic observations to measure the time it takes each moon to complete each orbit (the *period*). Using proportions and geometry, the radius of each orbit has also been determined. With these data in place, the speed of each moon in its orbit may be found by taking the circumference of each orbit ($2 \times \pi \times$ radius, assuming a circular orbit, with π, or *pi*, approximately equal to 3.14159) divided by the period. The acceleration of each moon in its respective orbit is equivalent to its centripetal acceleration, which is found by dividing the square of its speed by the radius of its orbit. Table 10.4 provides data for various moons in our solar system.

In an attempt to find the relationships in the data, the gravitational acceleration versus the inverse of the orbital radius squared was graphed in Figure 10.3.

TABLE 10.4

Central Planet Name	Central Planet Mass (earth masses)	Moon Name	Radius of Moon's Orbit (millions of m)	Period of Moon (days)	Orbital Speed of Moon (km/s)	Gravitational Acceleration of Moon in Orbit (m/s/s)
Earth	1	Moon	385	27.30	1.0	0.00273
Jupiter	318	Andrastea	129	0.30	31.5	7.67
Jupiter	318	Amalthea	181	0.50	26.5	3.87
Saturn	95	Atlas	138	0.60	16.6	2.01
Saturn	95	Epimetheus	151	0.69	15.8	1.66
Saturn	95	Mimas	186	0.94	14.3	1.11
Saturn	95	Enceladus	238	1.37	12.6	0.671
Saturn	95	Telesto	295	1.88	11.4	0.441
Saturn	95	Helene	378	2.74	10.0	0.267
Saturn	95	Rhea	527	4.52	8.4	0.137
Saturn	95	Hyperion	1,480	21.30	5.1	0.0173
Saturn	95	Iapetus	3,560	79.30	3.2	0.00299
Saturn	95	Phoebe	13,000	550.00	1.7	0.000226
Uranus	15	Miranda	129	1.40	6.7	0.343
Uranus	15	Ariel	191	2.52	5.5	0.159

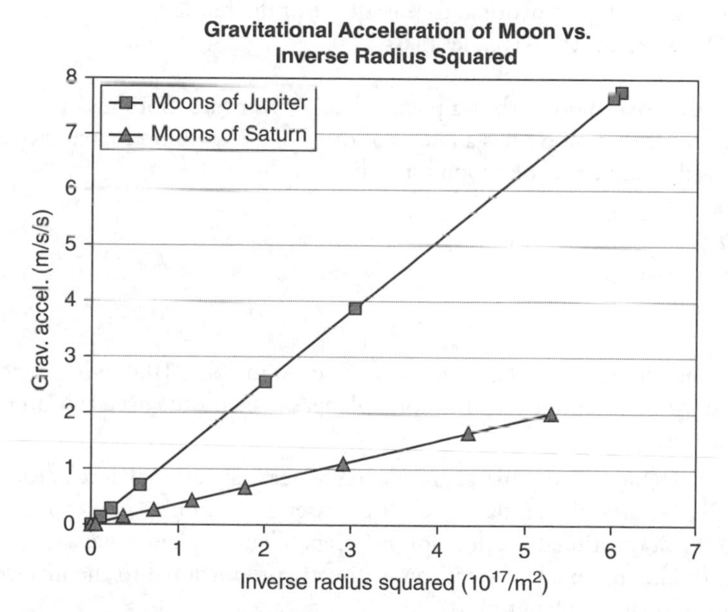

Figure 10.3

393. Which moon has a period closest in value to the period of Earth's moon?
(A) Andrastea
(B) Miranda
(C) Rhea
(D) Hyperion

394. Given just the data in the table, which two moons are most appropriate for studying the effect of central planet mass on gravitational acceleration?
(A) The moon and Atlas
(B) Andrastea and Miranda
(C) Atlas and Phoebe
(D) Iapetus and Arial

395. What is the best description of the relationship between radius of orbit and the period of the moon?
(A) The period is independent of the radius of orbit.
(B) The period has no consistent relationship with the radius of orbit.
(C) The larger the radius of orbit, the less the period.
(D) The larger the radius of orbit, the greater the period.

396. Figure 10.3 indicates that the gravitational acceleration of moons is:
(A) inversely proportional to the square of the radius.
(B) directly proportional to the radius.
(C) directly proportional to the square of the radius.
(D) independent of planet mass.

397. A particular moon orbits a planet that is 318 times more massive than Earth. If that moon has a radius of orbit of 422 million meters, what is a possible value for the moon's speed?
(A) 52 km/s
(B) 31 km/s
(C) 26 km/s
(D) 17 km/s

398. Using Figure 10.3, along with information in Table 10.4, what conclusion can be made about the effect of a planet's mass on the acceleration of its moons?
(A) Greater planet masses result in greater gravitational acceleration.
(B) Greater planet masses result in lesser gravitational acceleration.
(C) Gravitational acceleration is independent of planet mass.
(D) Gravitational acceleration is directly proportional to the inverse-square of planet mass.

399. According to the passage, which columns were calculated from data gathered by scientists?

(A) Period of moon orbit and central planet mass
(B) Central planet mass and radius of orbit
(C) Speed and gravitational acceleration
(D) Gravitational acceleration and period

400. About how many times does the moon orbit the earth while Phoebe completes one orbit of Saturn?

(A) 1.7
(B) 34
(C) 20
(D) 27

401. How far does Helene travel in 15 seconds?

(A) 10 km
(B) 41 km
(C) 150 km
(D) 2,400 km

402. If the moons of Mars (a planet with a mass 0.107 times that of Earth) were placed on the graph in Figure 10.3, where would they most likely appear?

(A) Above the line for the moons of Jupiter
(B) Below the line for the moons of Jupiter but above the line for the moons of Saturn
(C) Above the line for the moons of Jupiter initially, but then below that same line
(D) Below the line for the moons of Saturn

Passage 34

A sample of solid lauric acid was placed in a test tube and heated in a Bunsen burner in a chemistry classroom. The lauric acid melted. A thermometer was carefully placed in the test tube to measure the temperature of the sample. The burner was then turned off, and the sample was allowed to cool; the temperature was recorded in Figure 10.4. The warm liquid sample crystallized into a solid while the temperature was being recorded.

The *melting point* of a substance is the temperature at which that substance changes from the solid to the liquid phase. The *boiling point* of a substance is the temperature at which that substance changes from the liquid to the gaseous phase. Table 10.5 shows the boiling and melting points of other common substances.

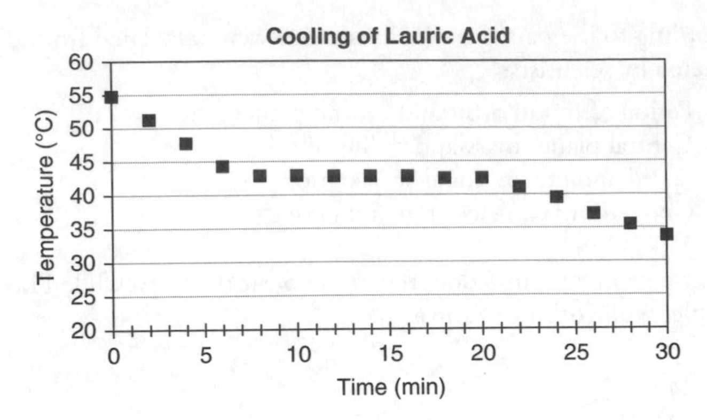

Figure 10.4

TABLE 10.5

Substance	Melting Point (°C)	Boiling Point (°C)
Water	0	100
Isopropyl alcohol	–90	82
Paraffin wax	59	343

403. Based on Figure 10.4, the temperature of the lauric acid sample decreased most rapidly during which time interval?

(A) 0 to 6 minutes
(B) 15 to 20 minutes
(C) 20 to 25 minutes
(D) 25 to 30 minutes

404. The thermometer was carefully placed in the test tube to measure the temperature of the lauric acid. If the thermometer was allowed to rest on the bottom of the test tube, what temperature may have been recorded at the start of the experiment (0 minutes)?

(A) 80°C
(B) 50°C
(C) 20°C
(D) 0°C

405. During the experiment, particles of lauric acid could either decrease in temperature or change from a liquid to a solid—they could not do both at the same time. In which time interval(s) did the particles' temperature decrease?

(A) 0 to 6 minutes
(B) 6 to 20 minutes
(C) 20 to 30 minutes
(D) Both 0 to 6 minutes and 20 to 30 minutes

406. What is the melting point of lauric acid?

(A) 55°C
(B) 43°C
(C) 34°C
(D) 0°C

407. What phase(s) of matter is present in the sample of lauric acid at 25°C?

(A) Only solid
(B) Only liquid
(C) Only gaseous
(D) Both solid and liquid

408. The sample of lauric acid was allowed to remain in the room, and the clock continued to run. If the trend found in Figure 10.4 continued, what would the stopwatch read when the sample reached room temperature of 25°C?

(A) About 32 minutes
(B) About 40 minutes
(C) About 50 minutes
(D) About 60 minutes

409. If twice the amount of lauric acid were heated in the test tube, at what temperature would one expect the lauric acid to crystallize into a solid?

(A) 43°C
(B) 86°C
(C) Somewhere between 43°C and 86°C
(D) A little lower than 43°C

410. What would most likely have occurred if this experiment had been conducted in an oven set at 50°C?

(A) The sample would have taken much longer to turn to a solid.
(B) The sample would have turned to a solid much more quickly.
(C) The sample would never have turned into a solid.
(D) The sample would not have decreased in temperature.

411. According to the data in Table 10.5, what phase(s) of matter is/are present in the sample of isopropyl alcohol at 25°C?

(A) Only solid
(B) Only liquid
(C) Only gaseous
(D) Both solid and liquid

412. The same experiment was conducted with a sample of water in a test tube. It was heated to 55°C and then removed from the heat and allowed to cool in the classroom. Which of the following graphs would most likely represent the results?

(A)

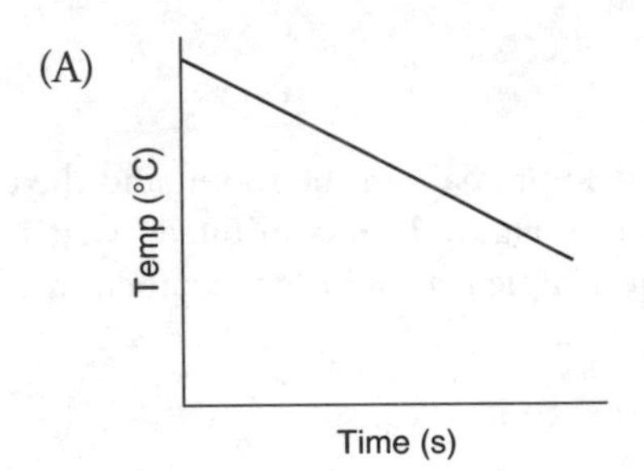

Figure 10.5

(B)

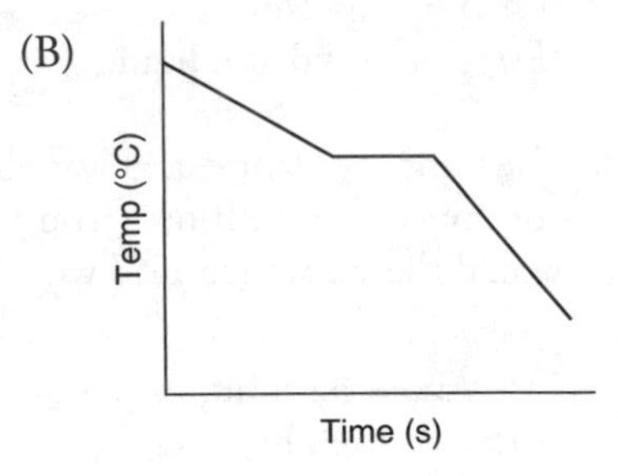

Figure 10.6

(C)

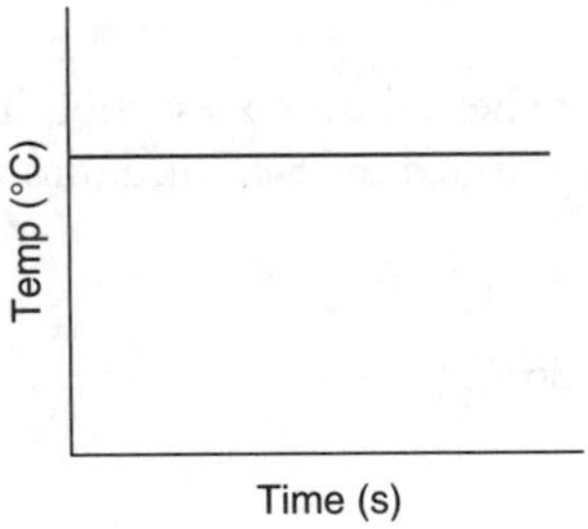

Temp (°C)

Time (s)

Figure 10.7

(D)

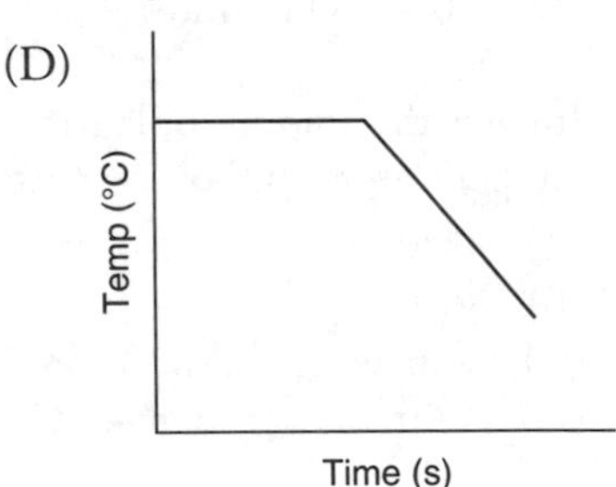

Figure 10.8

CHAPTER 11

Test 11

Passage 35

In the early 1800s, chemists started experimenting with different chemicals. They had the ability to measure the temperature, pressure, and mass of a sample of gas. In 1911, a chemist named Amadeo Avogadro published an observation that came to be known as Avogadro's law. This law states that any two gases that are held under the same pressure, temperature, and volume will contain the same number of molecules (measured in moles) regardless of the identity of the gases. Table 11.1 shows data collected from various gas samples at 1 atm and 0°C.

TABLE 11.1

Sample	Gas	Volume (L)	Mass of Gas (g)
1	Hydrogen (H_2)	11.2	1.0
2	Hydrogen (H_2)	5.6	0.5
3	Neon (Ne)	11.2	10.1
4	Neon (Ne)	22.4	20.2
5	Helium (He)	22.4	4.0
6	Helium (He)	44.8	8.0
7	Oxygen (O_2)	11.2	16.0
8	Oxygen (O_2)	5.6	8.0

413. According to Table 11.1, which sample of gas took up the most space?
(A) Sample 1
(B) Sample 4
(C) Sample 6
(D) Sample 7

414. Comparing Samples 1 and 3, one can state that:
(A) Samples 1 and 3 are the same size and the same mass.
(B) Sample 1 is larger, but Sample 3 is more massive.
(C) Sample 3 is larger, but Sample 1 is more massive.
(D) Samples 1 and 3 are the same size, but Sample 3 is more massive.

415. How does the number of gas molecules in Sample 4 compare to the number of gas molecules in Sample 5?
(A) Sample 4 has more molecules.
(B) Sample 5 has more molecules.
(C) Samples 4 and 5 have the same number of molecules.
(D) It is impossible to determine from the data given.

416. How much mass would a 22.4 L sample of oxygen gas have when measured at 1 atm and 0°C?
(A) 4.0 g
(B) 20.2 g
(C) 32.0 g
(D) 44.8 g

417. In the early 1800s, mass was measured with a double-pan balance. If an 11.2 L sample of neon is placed in one side of the balance, what volume of hydrogen would have to be placed on the other side to have an equal amount of mass?
(A) 224.0 L
(B) 112.0 L
(C) 22.4 L
(D) 11.2 L

418. Referring to Avogadro's law and Table 11.1, which of the following places the samples of gas in order from the least number of molecules to the most?
(A) Sample 1, Sample 3, Sample 7
(B) Sample 7, Sample 3, Sample 1
(C) Samples 3 and 1 have an equal number of molecules, but Sample 7 has more.
(D) Samples 1, 3, and 7 have an equal number of molecules.

419. What can be said about the mass of one molecule of helium and one molecule of oxygen gas?

(A) One molecule of helium has the same mass as one molecule of hydrogen gas.
(B) One molecule of oxygen is 8 times more massive than one molecule of helium.
(C) One molecule of helium is 8 times more massive than one molecule of oxygen gas.
(D) One molecule of oxygen gas is 4 times more massive than one molecule of helium.

420. Chemists counted molecules in the unit of *moles*. Avogadro stated that 1 mole (mol) of gas particles at 1 atm and 0°C takes up a volume of 22.4 L. What is the mass of 1 mol of oxygen gas molecules?

(A) 32.0 g
(B) 16.0 g
(C) 8.0 g
(D) 4.0 g

Passage 36

Photosynthesis is the process by which organisms such as green plants, algae, and cyanobacteria convert light energy from the sun to chemical energy stored in the bonds of carbohydrates. Chlorophyll is a pigment employed by many autotrophic organisms to absorb the various wavelengths of visible light from the sun for use in photosynthesis. A variety of photosynthetic pigments exist; they are specifically adapted for absorbing different ranges of the visible light spectrum and reflecting others. The absorption spectrum of chlorophyll and accessory pigments can be obtained through spectrophotometry and later used to gain insight into plant growth, determine the abundance of photosynthetic organisms in fresh- or saltwater, and evaluate water quality.

The data in Figure 11.1 and Table 11.2 were collected by students measuring the absorption spectra of three commonly encountered photosynthetic pigments.

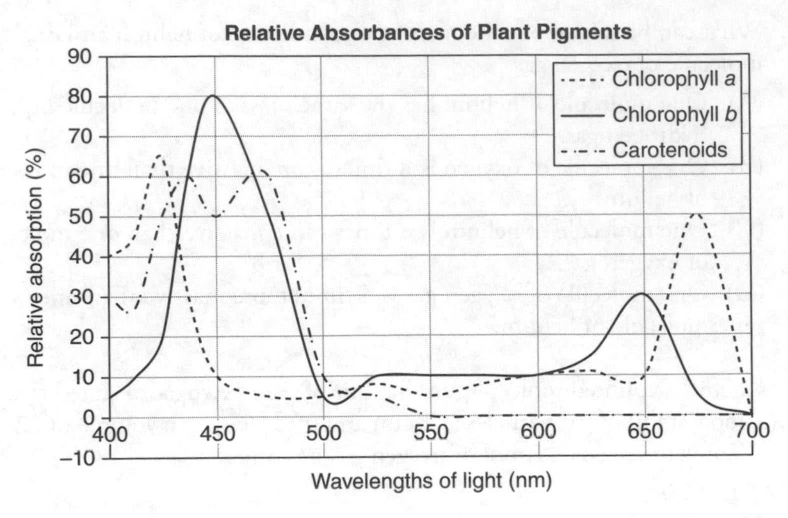

Figure 11.1

TABLE 11.2

Color	Wavelength
Red	620–750 nm
Orange	590–620 nm
Yellow	570–590 nm
Green	495–570 nm
Blue	450–495 nm
Violet	380–450 nm

421. Which of the following statements represents a valid assessment of the data?

(A) Chlorophyll *a* and *b* absorb the most green light.
(B) Neither chlorophyll *a*, *b*, nor the carotenoids absorb light in the wavelengths of 425 to 475 nm.
(C) Chlorophyll *a* and *b* have the most reflection in the wavelengths of 525 to 625 nm.
(D) Carotenoids absorb the most light in the red portion of the spectrum.

422. Using the information in Table 11.2, which of the following wavelengths of light would phycocyanin, a pale blue–colored accessory pigment, reflect most?

(A) 400 nm
(B) 460 nm
(C) 550 nm
(D) 750 nm

423. Based on Figure 11.1, which of the following wavelengths of visible light would be absorbed to promote the most photosynthetic activity in green plants?

(A) 400 nm
(B) 440 nm
(C) 550 nm
(D) 625 nm

424. Based on the information in Figure 11.1 and Table 11.2, which pigment has the highest relative absorbance in the red portion of the spectrum?

(A) Chlorophyll *a*
(B) Chlorophyll *b*
(C) Carotenoids
(D) All pigments absorb the same amount of light in this portion of the spectrum.

425. Which of the following statements about the relative absorbance levels of the pigments is most accurate?

(A) Chlorophyll *a* absorbs 8 times as much light as chlorophyll *b* at 450 nm.
(B) Carotenoids behave similarly to chlorophyll *a* and *b* at wavelengths greater than 550 nm.
(C) Chlorophyll *b* has the lowest relative absorbance of violet light.
(D) Carotenoids absorb approximately 60% as much light as chlorophyll *b* at 450 nm.

426. Which pigment would absorb the most violet light at a wavelength of 425 nm?

(A) Chlorophyll *a*
(B) Chlorophyll *b*
(C) Carotenoids
(D) All pigments absorb violet light equally.

427. Which wavelength represents the maximum absorbance of chlorophyll *b*?

(A) 425 nm
(B) 450 nm
(C) 515 nm
(D) 650 nm

428. From the information in the passage, one can conclude that chlorophyll appears green to the human eye because:

(A) wavelengths of light between 550 and 600 nm are highly absorbed.
(B) wavelengths in the green portion of the visible spectrum are absorbed.
(C) wavelengths in the green portion of the visible spectrum are reflected.
(D) wavelengths in the red portion of the visible spectrum are reflected.

429. Phycoerythrin is a photosynthetic pigment that is found in marine algae. It has absorption peaks at 495 nm and 566 nm and reaches its lowest values over 600 nm. Based on this information, one would expect phycoerythrin to:

(A) appear very similar to chlorophyll *a* and *b* to the naked eye.
(B) reflect green light and absorb red light.
(C) reflect red light and absorb green light.
(D) absorb red and green light equally.

Passage 37: Light Waves Versus Light Particles

In the late 1600s, scientists developed theories about the nature of light. Sir Isaac Newton theorized that light consisted of tiny particles. Christiaan Huygens, on the other hand, believed that light consisted of waves. What do we know about particles? Particles are small, localized objects that typically have certain physical properties like mass, color, or volume. They move in straight lines unless some outside force is acting on them.

Waves, on the other hand, consist of energy that moves through a medium (material). When waves hit a boundary from one material into another, some of that energy bounces back into the original material (reflection); some of the energy moves into the new material (refraction or transmission); and some of the energy transfers to thermal energy through heating (absorption). When waves encounter one another, they will increase in size and strength if similar parts of them overlap (constructive interference) and will decrease in size and strength if opposite parts of them overlap (destructive interference); after the interference, the waves continue moving in their original direction. Waves also spread out (diffract) when they hit a sharp edge or a tiny opening.

430. The front view in Figure 11.2 shows the image projected on a screen by the light from a tiny lightbulb placed in front of a square card. What can you conclude about the nature of light from the image shown?

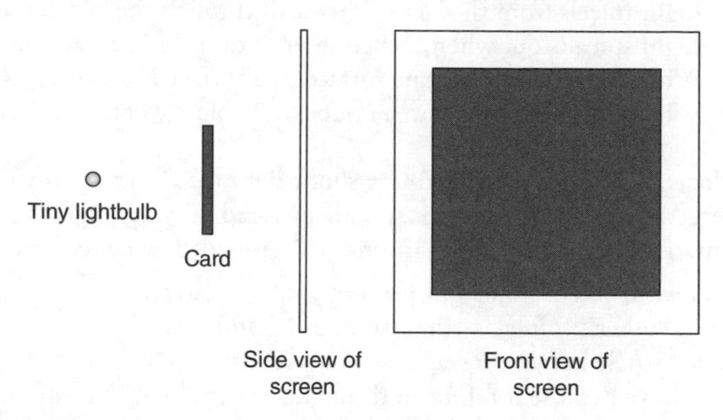

Figure 11.2

(A) Light behaves like a wave because it moves in all directions from the lightbulb.
(B) Light behaves like a wave because it does not move through the card.
(C) Light behaves like a particle because it moves in straight lines from the bulb to form the larger shadow on the screen.
(D) Light behaves like a particle because it reflects, refracts, and absorbs in this scenario.

431. Which of the following does NOT support Huygens's wave theory of light?
(A) A laser light reflects off a mirror and strikes the wall in the back of the lab.
(B) Light hits a black shirt and heats it up.
(C) Sunlight transmits through a thick pane of glass, illuminating a room.
(D) The smallest unit of light is a bundle of energy called a photon.

432. In the 1600s, it was believed that waves required a medium to travel. Which of the following observations did Huygens have difficulty explaining with his wave theory?

(A) Light travels from the sun to the earth through the vacuum of space.
(B) Light spreads out when it encounters a thin slit in a sheet of paper.
(C) When light hits an opaque surface, the surface heats up.
(D) White light separates into a rainbow of colors when it strikes a prism.

433. In the early 1800s, Thomas Young shone light through two tiny slits and observed an image on the screen that consisted of many regions of alternating bright and dark patterns. This provided evidence that light is:

(A) a wave because the light from one slit overlapped with the light from the other slit in areas that were strong and weak.
(B) a particle because it was able to make it through the two slits.
(C) a wave because it refracted through the slits, bent in multiple directions, and reflected and absorbed at the screen.
(D) multiple particles bouncing off the slits and moving to certain locations on the screen to make the bright areas and avoiding the dark areas.

434. When a laser beam of light strikes a penny, one might expect a sharp shadow to form. Instead, a circular, fuzzy-edged shadow forms with a bright spot in the middle. Which of the following best explains this phenomenon?

(A) Laser light particles move in straight lines, and some of those particles are blocked by the penny.
(B) The beam of laser light is a result of stimulated emission of electrons, and the electron particles are able to travel through the material of the penny.
(C) The light bounces off the edge of the penny and hits the middle of the shadow, causing the bright spot.
(D) The light diffracts around the sharp edge of the penny and interferes constructively at the center of the penny.

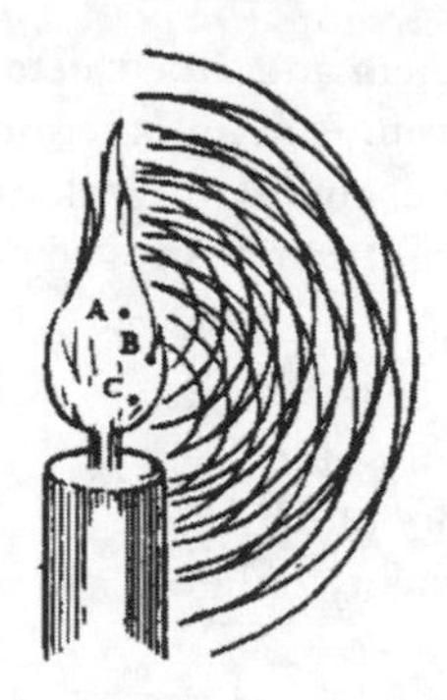

Figure 11.3

Source: Christiaan Huygens, Treatise on Light. *Translated by Silvanus P. Thompson.* The Project Gutenberg eBook, *http://www.gutenberg.org/files/14725/14725-h/14725-h.htm#Page_4.*

435. Figure 11.3 was most likely created by Huygens because:

(A) the circles show the pathways that particles take when they leave the candle flame.
(B) light is moving in all directions from the candle flame.
(C) the circles coming from Points A, B, and C are like ripples in a pond.
(D) light is coming from different points on the candle flame.

436. Is the following more likely to be attributed to Newton or Huygens and why?

If you consider the extreme speed with which light spreads on every side, as well as the fact that when it comes from different regions—even those directly opposite—the rays traverse one another without being hindered, you may well understand that when a person sees a luminous object, it cannot be by any transport of matter coming to that person from the object, in the same way that a shot or an arrow flies through the air. Therefore, light must spread in some other way, and that which can lead us to comprehend it is the knowledge we have of the spreading of sound in the air.

(A) Newton, because of the reference to the way light spreads
(B) Huygens, because it claims that light cannot travel like matter does
(C) Newton, because the emphasis is on the extreme speed of light
(D) Huygens, because it explains how we can see images of luminous objects

437. In the 1900s, Albert Einstein and others determined that the basic unit of light consists of a photon. Among other properties, this unit has a fixed amount of energy, depending on its location on the visible color spectrum. The fact that light can be broken down to fixed units is most consistent with:

(A) Newton's particle theory of light.
(B) Huygens's wave theory of light.
(C) Newton's wave theory of light.
(D) Huygens's particle theory of light.

438. A beam of direct sunlight moves along an axis through the middle of two polarizing filters, where the axis remains perpendicular to the filters. When the light passes through the first polarizing filter, its brightness reduces to 50%, regardless of its orientation. The light will remain at the same brightness after the second polarizing filter if that filter is aligned the same way as the first. If the second filter twists, however, the light passing through it will gradually be eliminated. Which of the following statements CANNOT be supported by this information?

(A) Because wave energy may be absorbed at a boundary, it is feasible that only 50% of the light energy passes through the first filter.
(B) Light particles may be blocked by objects, therefore it is feasible that only 50% of the light passes through the first filter.
(C) Polarizing filters may be used to control the intensity of light.
(D) The change in the light intensity as the second filter is rotated may be explained by the particle theory of light.

Passage 38

Two investigations were conducted on a sample of green nickel sulfate ($NiSO_4$) that was dissolved in water. The first investigation placed a sample of the nickel sulfate into a spectrometer. Most solutions absorb some wavelengths of light and allow other wavelengths to pass right through. The spectrometer changed the wavelength of light shining into the solution and recorded how much of the light was absorbed by the solution. An *absorbance* of 0.00 would indicate that all of the light shone into the solution passed through with no light being absorbed. Figure 11.4 shows the result of absorbance versus wavelength for a sample of $NiSO_{4(aq)}$.

The next experiment used five different solutions of $NiSO_{4(aq)}$ at different known concentrations (measured in molarity). Each solution was placed in a spectrometer set at 740 nm, and light was shone into each sample to determine the absorbance of each of the five solutions. Table 11.3 shows the results.

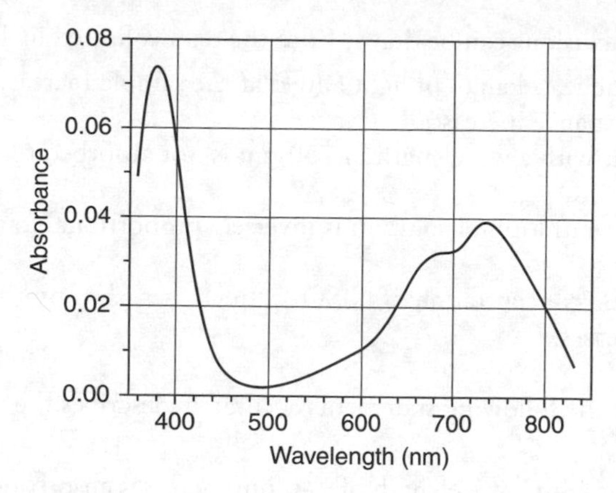

Figure 11.4

TABLE 11.3

Sample	Concentration of $NiSO_{4(aq)}$ (mol/L)	Absorbance
1	0.08	0.091
2	0.16	0.182
3	0.24	0.273
4	0.32	0.363
5	0.40	0.451

439. Which wavelength of light was absorbed to the greatest degree in the first investigation?

(A) 380 nm
(B) 500 nm
(C) 740 nm
(D) 830 nm

440. The wavelength of 490 nm is green light. Why is the absorbance of $NiSO_4$ low at 490 nm?

(A) The spectrometer is not accurate at the wavelength of 490 nm.
(B) The $NiSO_{4(aq)}$ is green in color, so green light from the spectrometer passes through the solution.
(C) The $NiSO_{4(aq)}$ is green in color, so the green light from the spectrometer is absorbed to a large extent.
(D) The green light gets reflected back to the spectrometer when it contacts the green $NiSO_{4(aq)}$ solution.

441. What conclusions can be drawn from the data collected in Table 11.3?

(A) As the wavelength of light aimed at the sample increases, the absorbance increases.
(B) Light with a wavelength of 740 nm is not absorbed to a great extent by $NiSO_{4(aq)}$.
(C) Concentration of solution is inversely proportional to absorbance of light.
(D) As the concentration of $NiSO_{4(aq)}$ increases, the absorbance of light increases.

442. Which of the following statements accurately describes the trend found in Table 11.3?

(A) Light with a wavelength of 740 nm increases absorbance by about 0.09.
(B) Doubling the concentration of $NiSO_{4(aq)}$ causes an increase in absorbance of about 0.90.
(C) An increase of 0.08 mol/L causes an increase in the absorbance of about 0.90.
(D) The absorbance of $NiSO_{4(aq)}$ stops increasing at a concentration of 0.40 mol/L.

443. Which set of data would best represent the results if Investigation 2 were repeated with a wavelength of 490 nm instead of 740 nm?

	Sample	**Concentration $NiSO_{4(aq)}$ (mol/L)**	**Absorbance**
(A)	1	0.08	0.091
	2	0.16	0.182
	3	0.24	0.273
	4	0.32	0.363
(B)	1	0.08	0.049
	2	0.16	0.098
	3	0.24	0.147
	4	0.32	0.196
(C)	1	0.08	0.003
	2	0.16	0.006
	3	0.24	0.009
	4	0.32	0.012
(D)	1	0.08	0.25
	2	0.16	0.51
	3	0.24	0.76
	4	0.32	0.99

444. The process of spectrometry works well on colored solutions such as green nickel sulfate ($NiSO_4$) and cupric sulfate ($CuSO_4$). Why might spectrometry not work well on solutions such as table salt dissolved in water ($NaCl_{(aq)}$)?

(A) Table salt is clear, so it will absorb all wavelengths of light.
(B) Table salt is clear, so all colors of light will pass through the solution.
(C) The researcher cannot vary the concentration of table salt.
(D) Spectrometry is not appropriate for food-grade substances.

445. Experiment 2 was repeated with a sixth sample of $NiSO_4$ solution with a concentration of 0.48 mol/L. However, the test tube had fingerprints on the glass where the light passed through. The fingerprints absorbed some light from the spectrometer. What would be an expected value for the absorbance?

(A) 0.451
(B) 0.522
(C) 0.542
(D) 0.622

446. What would be the results of Investigation 1 if a more concentrated solution of $NiSO_{4(aq)}$ were used?

(A)

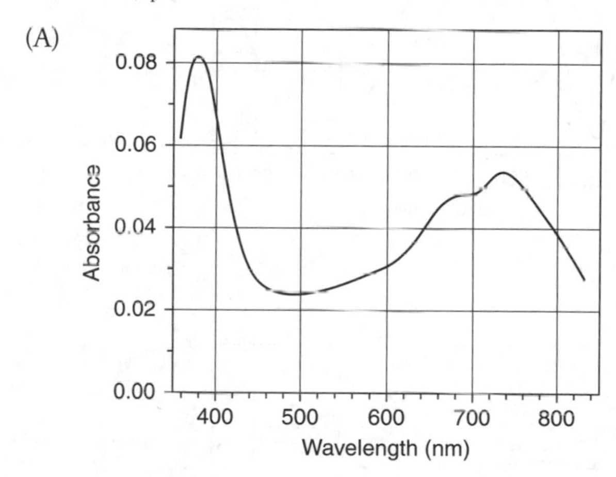

Figure 11.5

(B)

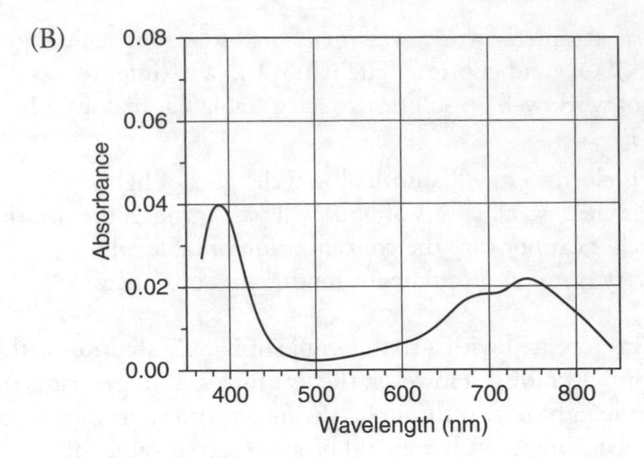

Figure 11.6

(C)

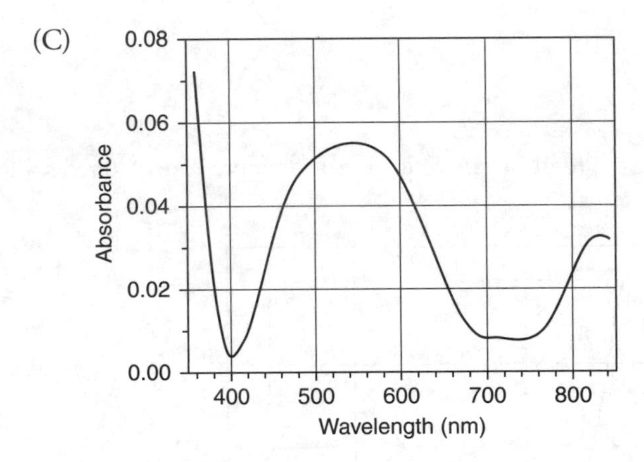

Figure 11.7

(D)

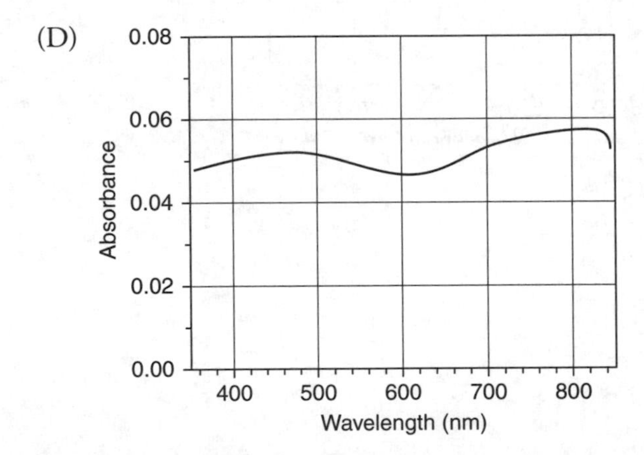

Figure 11.8

Passage 39: Spectral Analysis of Bacterial Growth

Staphylococcus aureus (*S. aureus*) is a bacterium found on the skin of 25% of healthy individuals with no adverse effects. However, when transferred to food products, the toxins it can produce have been known to cause food-borne illness, particularly in cooked and cured meats. It is important to determine conditions that promote the growth of pathogenic species of bacteria and apply that knowledge in food preparation and handling. Two groups of students set out to determine the ideal growth conditions for *S. aureus*. In each group, bacteria were inoculated in a nutrient broth and allowed to grow over a period of time.

When analyzing the growth of bacteria in a liquid medium, an increase in turbidity corresponds to an increase in the bacteria growing in suspension. Because the bacterial cells scatter light, spectrophotometry can be used to determine bacterial growth. Changes in the logarithmic absorbance scale on the spectrophotometer correspond to changes in the number of cells, and a growth curve can be plotted by graphing the absorbance readings from the spectrophotometer versus time. The rate of growth can be determined by the slope of the lines, and the type of growth occurring at a given time can be determined by the shape of the curve.

Group 1

The students in Group 1 set out to determine whether temperature affected the growth rate of *S. aureus* in a nutrient-rich medium. They tested *S. aureus* by inoculating a flask of nutrient-rich broth with a pH of 6 at the following temperatures: 3°C, 20°C, 37°C, 45°C, and 60°C. The spectrophotometer absorbance readings are shown in Table 11.4.

TABLE 11.4

Time (min)	Trial 1 (3°C)	Trial 2 (20°C)	Trial 3 (37°C)	Trial 4 (45°C)	Trial 5 (60°C)
0	0.011	0.011	0.011	0.012	0.013
48	0.011	0.030	0.022	0.045	0.013
62	0.011	0.067	0.035	0.081	0.013
68	0.011	0.072	0.047	0.056	0.013
82	0.011	0.085	0.070	0.075	0.013
88	0.011	0.090	0.081	0.078	0.013
97	0.011	0.098	0.099	0.078	0.013
108	0.011	0.108	0.109	0.079	0.013
118	0.011	0.117	0.137	0.088	0.013
128	0.011	0.126	0.149	0.089	0.013
138	0.011	0.135	0.167	0.098	0.013
148	0.011	0.144	0.174	0.106	0.013

(*Continued*)

TABLE 11.4 (*Continued*)

Time (min)	Trial 1 (3°C)	Trial 2 (20°C)	Trial 3 (37°C)	Trial 4 (45°C)	Trial 5 (60°C)
157	0.011	0.152	0.190	0.109	0.013
168	0.011	0.162	0.207	0.114	0.013
180	0.011	0.173	0.222	0.117	0.013
188	0.011	0.180	0.297	0.118	0.013
198	0.011	0.189	0.288	0.126	0.013
600	0.011	0.551	0.800	0.125	0.013
1,470	0.011	0.553	0.800	0.130	0.013

Group 2

Students in Group 2 planned to determine the effect of pH on the growth of *S. aureus*. They wanted to test growth at a range of values from highly acidic (low pH) to highly basic/alkaline (high pH). The cultures were grown at a temperature of 37°C, and the pH levels of each culture were as follows: 3, 5, 6, 7, and 9. Table 11.5 shows their data.

TABLE 11.5

Time (min)	Trial 1 (pH 3)	Trial 2 (pH 5)	Trial 3 (pH 6)	Trial 4 (pH 7)	Trial 5 (pH 9)
0	0.012	0.012	0.011	0.011	0.013
48	0.012	0.026	0.022	0.045	0.013
62	0.012	0.031	0.035	0.054	0.013
68	0.012	0.032	0.047	0.059	0.013
82	0.012	0.037	0.070	0.068	0.013
88	0.012	0.038	0.081	0.073	0.013
97	0.012	0.041	0.099	0.079	0.013
108	0.012	0.044	0.109	0.087	0.013
118	0.012	0.047	0.137	0.094	0.013
128	0.012	0.050	0.149	0.101	0.013
138	0.012	0.053	0.167	0.108	0.013
148	0.012	0.056	0.174	0.115	0.013
157	0.012	0.059	0.190	0.121	0.013
168	0.012	0.062	0.207	0.129	0.013
180	0.012	0.066	0.222	0.137	0.013

(*Continued*)

TABLE 11.5 (*Continued*)

Time (min)	Trial 1 (pH 3)	Trial 2 (pH 5)	Trial 3 (pH 6)	Trial 4 (pH 7)	Trial 5 (pH 9)
188	0.012	0.068	0.297	0.143	0.013
198	0.012	0.071	0.288	0.150	0.013
600	0.012	0.192	0.800	0.431	0.013
1,470	0.012	0.195	0.851	0.440	0.013

447. Which of the following is a valid assessment of the data from Group 1?
(A) As temperature increases, the growth rate of *S. aureus* increases proportionally.
(B) *S. aureus* grows optimally at a temperature of 37°C.
(C) As temperature decreases, so does the growth rate of *S. aureus.*
(D) *S. aureus* exhibits exponential growth at a temperature of 37°C.

448. The stationary phase of growth is entered as the nutrients in the medium begin to run out and the growth of bacteria changes the conditions in the flask. Cell division slows, and the turbidity ceases to increase because the overall population remains unchanged. When did the bacteria in Group 1/Trial 3 likely begin the stationary phase of growth?
(A) 97 minutes
(B) 148 minutes
(C) 600 minutes
(D) 1,470 minutes

449. Which of the following statements is most accurate?
(A) *S. aureus* responds well to increases in pH above 7.
(B) pH factor does not have an effect on the growth of *S. aureus.*
(C) *S. aureus* is viable at pH levels below 2.
(D) pH is a control for Group 1 and an experimental variable for Group 2.

450. *Generation time* is the amount of time it takes for the bacterial population to double. Some bacteria, such as *E. coli*, have a doubling time of 20 minutes under ideal conditions, while other bacteria may take days to double their population size. Which of the following is most likely the approximate generation time for the bacteria in Group 1/Trial 3 during the first 100 minutes of the experiment?
(A) 6 minutes
(B) 29 minutes
(C) 75 minutes
(D) 100 minutes

451. Which of the following does NOT represent a controlled variable for both groups?

(A) The nature of the nutrient media
(B) The temperature of incubation
(C) The strain of *S. aureus* used
(D) The wavelength of light set on the spectrophotometer

452. Which of the following is a valid assessment of the data in Table 11.5?

(A) *S. aureus* has a faster growth rate at pH 7 than at pH 5.
(B) *S. aureus* achieves a higher overall turbidity at pH 5 than at pH 7.
(C) *S. aureus* achieves maximum growth at pH levels below 4.
(D) *S. aureus* has a faster growth rate at pH 5 than at pH 7.

453. The data for Group 2/Trial 5 indicate that:

(A) at this high level of acidity, *S. aureus* bacteria cannot grow.
(B) at this high level of alkalinity, *S. aureus* bacteria cannot grow.
(C) at this low level of alkalinity, *S. aureus* bacteria cannot grow.
(D) at this high level of alkalinity, *S. aureus* bacteria thrive.

454. Which of the following trials were run under the same conditions for Groups 1 and 2?

(A) Group 1/Trial 1 and Group 2/Trial 1
(B) Group 1/Trial 2 and Group 2/Trial 4
(C) Group 1/Trial 3 and Group 2/Trial 3
(D) Group 1/Trial 5 and Group 2/Trial 5

455. If scientists wanted to do further testing on the growth of *S. aureus* to inform food processing and handling decisions, such as the effect of salinity, what conditions would be best as the controls for their test?

(A) 37°C and pH 6
(B) 37°C and pH 5
(C) 20°C and pH 6
(D) 45°C and pH 7

456. Which of the following graphs most closely resembles the shape of a graph that could be drawn for Group 1/Trial 3?

(A)

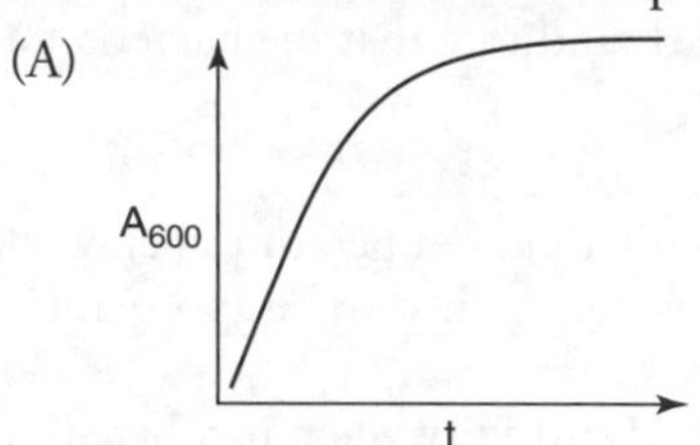

Figure 11.9

(B)

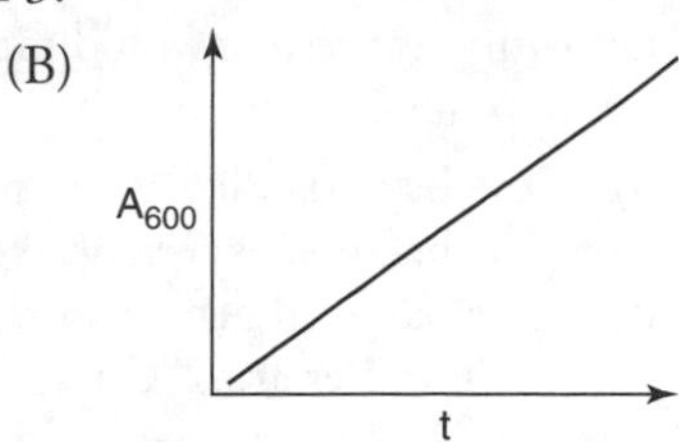

Figure 11.10

(C)

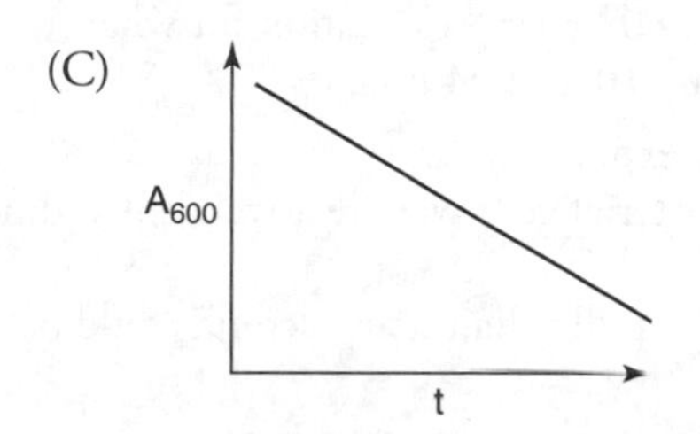

Figure 11.11

(D)

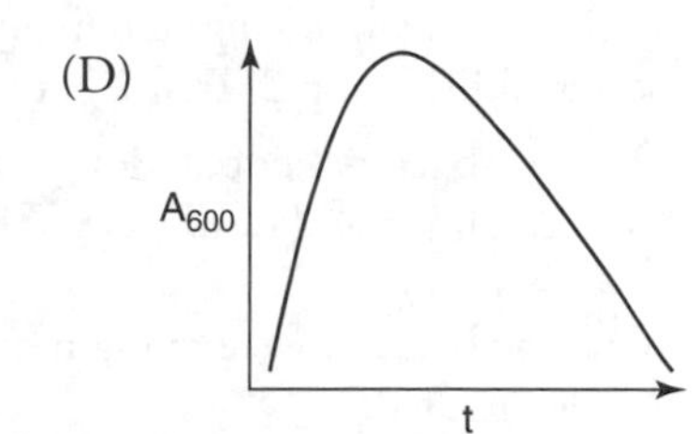

Figure 11.12

457. Chemical inhibitors such as sodium benzoate are often used as food preservatives because of their ability to retard, although not completely inhibit, bacterial growth. If a study were done on the effects of sodium benzoate on the growth of *S. aureus* at 37°C, one would expect to see results similar to those of:

(A) Group 2/Trial 1.
(B) Group 1/Trial 5.
(C) Group 2/Trial 2.
(D) Group 1/Trial 3.

458. What is the independent variable for both Group 1 and Group 2?

(A) Time
(B) pH
(C) Temperature
(D) The type of bacteria being tested

459. Meat in a delicatessen tested positive for the presence of bacteria, and this bacteria was to be identified using spectrophotometry. Which of the following test results would most clearly indicate that the bacteria found was *S. aureus*?

(A) The bacteria died at temperatures above 55°C.
(B) The bacteria experienced rapid growth at temperatures below 20°C.
(C) The bacterial generation time was determined to be approximately 30 minutes at 37°C.
(D) The bacteria showed no increase in turbidity when incubated in an ice bath.

460. Which of the following statements is the best explanation for what occurred in Group 2/Trial 2 between 10 and 24 hours?

(A) The bacterial population grew exponentially.
(B) The birth and death rate of bacterial cells were relatively equal during this time.
(C) The bacteria were dying more rapidly than new bacteria could be generated.
(D) No new bacteria were generated or died during this time.

CHAPTER 12

Timed Test

On an ACT science test, you'll have 35 minutes to read seven passages and answer a total of 40 questions. This chapter is a complete ACT practice test and is designed to help you practice your pacing on the test. So set your timer, and let's get started!

Passage 40

In 1922, Niels Bohr revised the atomic model to include a positively charged nucleus surrounded by negatively charged electrons that traveled in well-defined shells around the nucleus. The shells can be thought of as concentric circles around the nucleus. A neutral atom contains the same number of protons in the nucleus as electrons surrounding the nucleus. The inside shell can hold two electrons, and the second shell can hold eight electrons. An electrostatic attraction occurs between the positively charged protons in the nucleus and the negatively charged electrons. A representation of Bohr's shell model is shown in Figure 12.1.

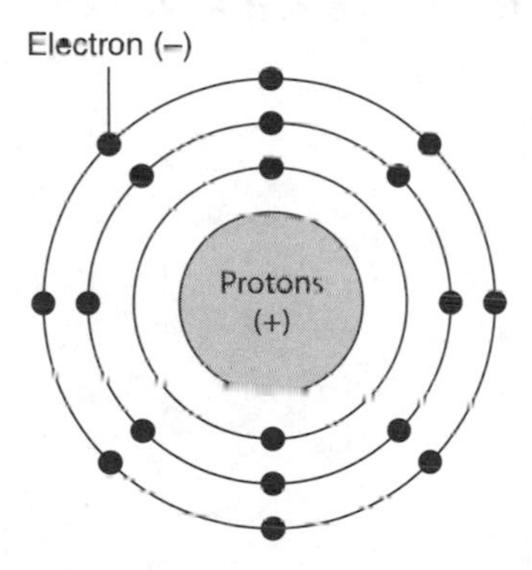

Figure 12.1

Partial evidence for this atomic shell model comes from the study of ionization energies of different elements. *Ionization energy* is the amount of energy required to remove an electron from an atom or ion in the gaseous state.

The first ionization energy removes the electron farthest from the nucleus of a neutral atom and can be represented by the following formula:

$$X + \text{Ionization energy} \rightarrow X^{+1} + \text{Electron}^{-1}$$

where X represents a neutral atom.

The nth ionization energy removes additional electrons from an already charged ion. For example, the third ionization energy can be represented by the following formula:

$$X^{+2} + \text{Ionization Energy} \rightarrow X^{+3} + \text{electron}^{-1}$$

The ionization energies in kJ/mol of the first 10 elements can be found in Table 12.1.

461. Which of the following statements regarding the ionization energy of the third electron of the elements listed is true?

(A) There is a general and consistent increase as the elements get larger.
(B) There is a general and consistent decrease as the elements get larger.
(C) Increases are then followed by a decrease.
(D) After a small increase and large dip, there is a gradual increase.

462. How much energy is required to remove an electron from nitrogen as shown in the following equation?

$$N^{+3} + \text{Ionization energy} \rightarrow N^{+4} + \text{Electron}^{-1}$$

(A) 1,402 kJ/mol
(B) 2,856 kJ/mol
(C) 4,578 kJ/mol
(D) 7,475 kJ/mol

TABLE 12.1

Atom	Protons	First	Second	Third	Fourth	Fifth	Sixth	Seventh	Eighth	Ninth	Tenth
H	1	1,312									
He	2	2,372	5,250								
Li	3	520	7,298	11,815							
Be	4	900	1,757	14,849	21,007						
B	5	801	2,427	3,659	25,025	32,827					
C	6	1,087	2,353	4,621	6,223	37,831	47,277				
N	7	1,402	2,856	4,578	7,475	9,445	53,267	64,360			
O	8	1,314	3,388	5,300	7,469	10,990	13,327	71,330	84,078		
F	9	1,681	3,374	6,050	8,408	11,023	15,164	17,868	92,038	10,6434	
Ne	10	2,081	3,952	6,122	9,371	12,177	15,238	19,999	23,070	115,380	131,432

463. Figure 12.2 shows an electron being removed from the element oxygen. How much energy is associated with the image shown?

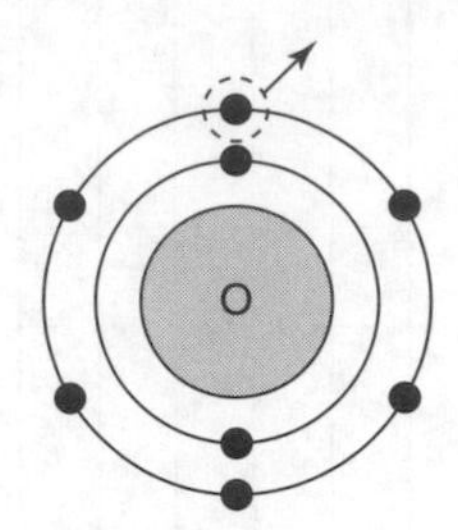

Figure 12.2

(A) 1,314 kJ/mol
(B) 3,388 kJ/mol
(C) 71,330 kJ/mol
(D) 84,078 kJ/mol

464. Referring to Table 12.1, what evidence supports the fact that each element has only two electrons in the first, innermost shell?

(A) The first two ionization energies are always smaller than the rest.
(B) The last two ionization energies are always significantly larger than the rest.
(C) There is a significant jump between the second and third ionization energies for most elements.
(D) The first two ionization energies of helium and lithium are both relatively small.

465. How many kJ of energy would be required to remove *all* of the electrons from 1 mol of helium atoms?

(A) 2,372 kJ
(B) 5,250 kJ
(C) 7,622 kJ
(D) 10,500 kJ

Passage 41

Many companies advertise that their brand of battery outlasts the competitors' batteries. A series of experiments were conducted to compare batteries from different manufacturers. Figure 12.3 shows the results of tests of four different brands of batteries. Two AA batteries were tested in an incandescent bulb flashlight, and the combined voltage was tested for continuous use over time. The flashlight operated effectively for voltage values greater than 2.2 V.

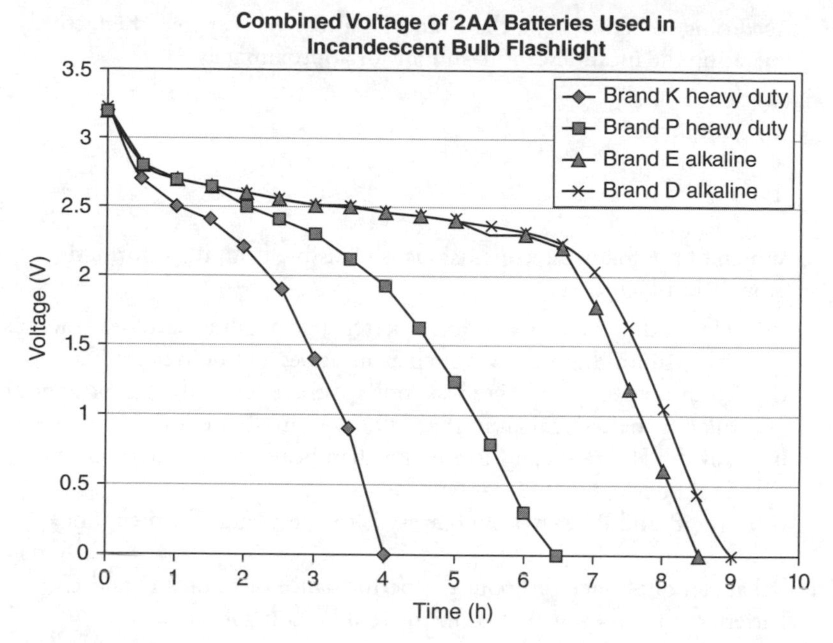

Figure 12.3

Table 12.2 compares how Brands D and E alkaline batteries performed in different devices. The table displays the time the combined voltage of two batteries remained above 2.2 V. The voltages were checked every quarter-hour.

TABLE 12.2

	Time two AA alkaline batteries exceeded 2.2 V (hours)					
Device	**Brand D Trial 1**	**Brand D Trial 2**	**Brand D Trial 3**	**Brand E Trial 1**	**Brand E Trial 2**	**Brand E Trial 3**
Incandescent bulb flashlight	7.00	6.75	6.75	6.75	6.50	6.50
LED bulb flashlight	24.25	25.50	24.00	23.75	25.75	24.75
Remote-control toy car	2.00	8.50	6.50	14.25	4.50	8.50
Plug-and-play video game	11.50	12.50	13.50	13.50	12.25	13.25

466. According to Figure 12.3, the Brand P battery was capable of effectively operating the incandescent flashlight for approximately:

(A) 0.5 hours.
(B) 2.0 hours.
(C) 3.5 hours.
(D) 6.5 hours.

467. Which of the following conclusions is plausible from the information provided in Figure 12.3?

(A) Heavy-duty batteries perform better than alkaline batteries when used in medium-drain devices such as incandescent bulb flashlights.
(B) In the experiment, there was more consistency in the performance of alkaline batteries than in that of heavy-duty batteries.
(C) Alkaline batteries perform better than heavy-duty batteries in high-drain devices such as camera flashes.
(D) The Brand P heavy-duty battery is the best value for the money spent.

468. What can one conclude about the performance of Brands D and E batteries when they were used in the remote-control car?

(A) Brand E performed significantly better than Brand D.
(B) Brand D performed significantly better than Brand E.
(C) The two brands' performances were approximately the same.
(D) The variability of the data prevents any valid comparison.

469. Which of the following statements is true according to Figure 12.3?

(A) In the first few hours of testing, the voltage drop for Brand D was greater than that for Brand K.
(B) Brand P's performance over the first four hours was nearly identical to that of Brand E.
(C) At voltages under 2.2 V, the voltage of Brand E drops more rapidly than that of Brand P.
(D) Brand K lasts longer than Brand P.

470. Which of the following steps in the experimental procedure is most important in the comparison of battery performance between brands?

(A) The batteries tested were manufactured in the same year.
(B) The batteries tested were not previously used in any other device.
(C) The temperature was controlled in the testing room.
(D) The time intervals for measuring voltages were the same for each brand.

471. Approximately how much longer than the incandescent bulb flashlight does the LED bulb flashlight last?

(A) Twice as long
(B) Four times as long
(C) Six times as long
(D) Twenty-four times as long

Passage 42

In the early 1900s, scientists began to suspect that unseen matter exists in our universe. The term "dark matter" was subsequently coined for matter that does not emit or reflect light. In the 1970s, astronomer Vera Rubin studied the rotation of stars within galaxies. Her data showed a discrepancy between the predicted orbital speeds of the stars and their measured periods. Figure 12.4 shows sample data for Galaxy UGC 11748 similar to the data that Rubin gathered. The predicted values of orbital speeds are calculated from Newton's universal law of gravitation using the amount of visible mass in the galaxy. The measured speeds are based on experimental observations of orbital periods of the stars.

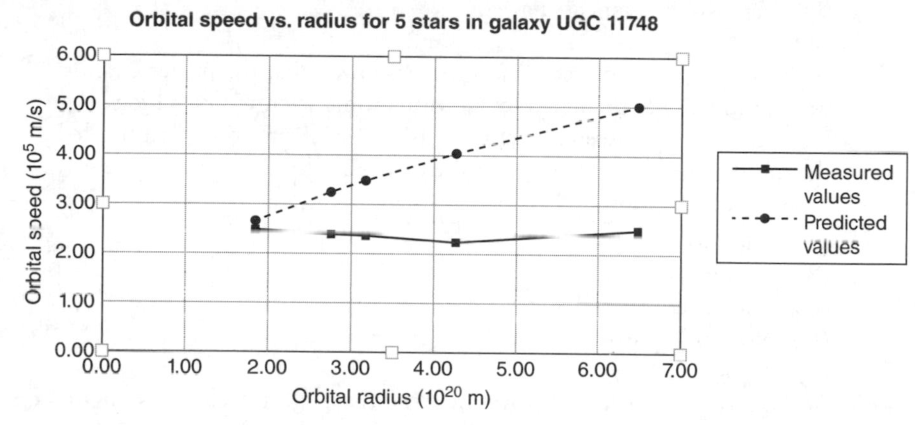

Figure 12.4

Two scientists are engaged in a debate about the interpretation of these results. Here are their opening remarks:

Scientist 1

The discrepency between the measured and predicted values for obital speeds of stars in distant galaxies clearly points to the existence of dark matter. By assuming additional, unseen mass to the galaxies, Newton's law of gravitation works perfectly to reconcile the discrepancy. Dark matter also helps explain the vast abundance of galaxies in our universe as well as intricate patterns of the cosmic microwave background radiation. Exciting research to detect dark matter particles is being conducted around the world.

Scientist 2

Many years of research has not been able to find the elusive "dark matter" particle. Why should we hang our hat on this theory? Remember that in the early 1900s, Newton's theory of gravity was not sufficient to explain the complex motion of Mercury. Some scientist proposed the existence of "Vulcan," an unseen planet, but that "dark matter" planet was never found. Rather, Einstein's general theory of relativity was needed to modify the Newtonian analysis to make sense of the observations of Mercury. I believe that a modification of Einstein's general theory of relativity is needed to explain Vera Rubin's observations.

472. What statement would both scientists agree upon?

(A) The dark matter particle has not yet been detected.
(B) Dark matter exists throughout our universe.
(C) Current theories of gravity must be modified to explain Rubin's observations.
(D) Vera Rubin's data are flawed.

473. In the 1820s, astronomical observations showed that the planet Uranus was not moving as predicted by Newton. Some scientists concluded that there must be an unseen planet responsible for this motion, and this led to the subsequent discovery of Neptune in 1846. This information best supports the position of:

(A) Scientist 1.
(B) Scientist 2.
(C) Both scientists.
(D) Neither scientist.

474. At the turn of the century, it was believed that light traveled in straight lines. However, in 1907, Einstein predicted that light could noticeably bend in the presence of large gravitational fields. This bending was experimentally detected during a solar eclipse in 1919 by the distortion of star position. The bending of light around galaxies is more pronounced than expected based on only the visible mass observed in those galaxies. This phenomenon, known as gravitational lensing, best supports the position of:

(A) Scientist 1.
(B) Scientist 2.
(C) Both scientists.
(D) Neither scientist.

475. Scientist 2 believes that the difference in the two lines in Figure 12.4 is most likely due to:

(A) the presence of unseen matter in the galaxy.
(B) the inadequacy of current theories of gravity.
(C) flaws in Vera Rubin's observation of the stars in the galaxy.
(D) the existence of a cosmic microwave background radiation.

476. According to experimental observations of the five stars in Galaxy UGC 11748, the orbital speed of the stars:

(A) increases significantly with orbital radius.
(B) decreases significantly with orbital radius.
(C) remains relatively unchanged with orbital radius.
(D) increases and then decreases with orbital radius.

477. Vera Rubin once stated, "If I could have my pick, I would like to learn that Newton's laws must be modified in order to correctly describe gravitational interactions at large distances. That's more appealing than a universe filled with a new kind of sub-nuclear particle." This statement most closely aligns with the arguments of:

(A) Scientist 1.
(B) Scientist 2.
(C) Both scientists.
(D) Neither scientist.

478. Which of the following would best support Scientist 1's position?

(A) The calculated values for orbital speed in Figure 12.4 increase with orbital radius.
(B) Thomas Kuhn's 1962 book *The Structure of Scientific Revolutions* argues that anomalies in scientific observations lead to periods of revolutionary science where old theories are superseded by new theories.
(C) In the vacuum of space, the gravitational field is inversely proportional to the square of the distance from the center of a mass.
(D) It took almost 50 years between the initial papers predicting the existence of the subatomic particle called the Higgs boson and its discovery in 2013.

Passage 43

When a mass hanging from a spring is pulled down and released, it will vibrate with a regular rhythm. The time it takes a spring to vibrate back and forth through one cycle is defined as the "period." Three student groups conduct separate experiments that attempt to measure how the period of a spring's vibration is affected by three variables. Student Group A varies the mass hanging from the spring, but uses the same spring each time and stretches the spring the same amount before releasing it. Student Group B uses different springs, but uses the same mass and stretches the spring the same amount before releasing it. Student Group C uses the same spring and same mass each time, but stretches the spring different amounts each time. The time for one cycle (period) was measured for each. The data for each experiment are shown in Table 12.3.

TABLE 12.3 Group A—Effect of Mass on Period

Trial #	Mass (kg)	Period (s)
1	0.00	0.00
2	1.00	1.35
3	2.00	1.85
4	3.00	2.35
5	4.00	2.70
6	5.00	3.00

TABLE 12.4 Group B—Effect of Spring Stiffness on Period

Trial #	Stiffness (N/m)	Period (s)
1	5.0	9.00
2	10.0	6.30
3	15.0	5.10
4	20.0	4.50
5	25.0	4.00
6	30.0	3.60

TABLE 12.5 Group C—Effect of Stretch Distance on Period

Trial #	Stretch (m)	Period (s)
1	0.10	3.98
2	0.15	4.00
3	0.20	3.99
4	0.25	4.01
5	0.30	4.01
6	0.35	4.03

479. According to Group A's data and the information in the passage, how much time would it take a 2.00 kg mass to vibrate through 10 complete cycles?

(A) 1.85 seconds
(B) 3.50 seconds
(C) 10.00 seconds
(D) 18.50 seconds

480. Which of the following is not a controlled variable in Group B's experiment?

(A) Mass
(B) Stretch distance
(C) Gravitational field
(D) Spring stiffness

481. Which of the following statements best describes the effect of mass on the period of vibration of a spring?

(A) The period increases linearly with mass.
(B) Mass has no significant effect on the period of vibration.
(C) The period increases as mass increases, with greater gains for smaller masses.
(D) The period increases as mass increases, with greater gains for larger masses.

482. Assuming Group B and Group C used the same mass in both experiments, what spring stiffness did Group C most likely use in their experiment?

(A) 10.0 N/m.
(B) 15.0 N/m.
(C) 20.0 N/m.
(D) 25.0 N/m.

483. Using the same mass as Group B, which of the following is the best predicted value for the vibration period of a spring with a stiffness of 40 N/m?

(A) 2.8 seconds.
(B) 3.2 seconds.
(C) 3.6 seconds.
(D) 4.0 seconds.

484. Which of the following is the best conclusion statement for all three experiments?

(A) Mass and spring stiffness both affect the period of oscillation, but stretch distance has virtually no effect.
(B) Mass and spring stiffness both increased the period of oscillation, but stretch distance only increased the period slightly.
(C) All three variables had no significant effect on the period of oscillation.
(D) All three variables had a significant effect on the period of oscillation.

Passage 44

Named after Johannes Kepler, the Kepler satellite telescope is part of a project that seeks to answer a question that has been asked since humanity first looked at the stars: "Does life exist somewhere else in the universe?" One of the goals of the Kepler project is to determine the abundance of terrestrial (Earth-like) planets that would be capable of harboring life. To accomplish this goal, the Kepler telescope monitors over 100,000 stars with similar characteristics to our Sun.

To support life, it is assumed that a planet must have water in liquid form, and this depends on the planet's surface temperature. Two key factors affect these surface temperatures. First, the temperature of the host star must be between 4500 K and 7000 K. The second and most important determining factor is the distance at which the planet orbits the star. A planet is in the "habitable zone," or life zone, if it is within a certain distance from its host star. Figure 12.5 shows the inner radius (dashed line) and outer radius (solid line) of the habitable zone of stars similar to our sun. The radii are measured in astronomical units (AUs); one AU is the distance between the Earth and the Sun.

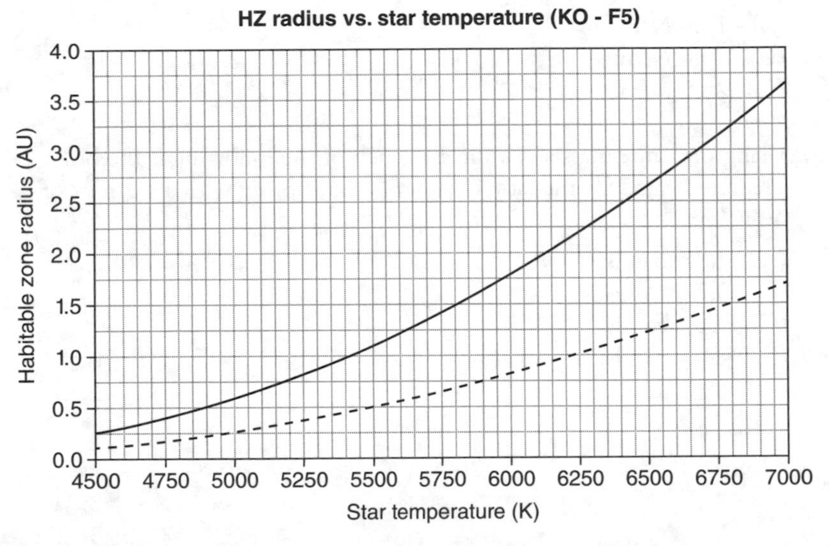

Figure 12.5

Since planets are not luminous sources, it's very difficult to see them in distant solar systems. To identify possible planets, the Kepler satellite watches for when the planet crosses in front of the star, an event called transit. When a planet transits, a percentage of light from the star is blocked. A minimum of three transits at regular intervals are necessary before declaring a planet candidate.

The transit light curve graph in Figure 12.6 shows two planets orbiting a star with an effective surface temperature of 6250 K. Kepler 426-B is shown as a dashed line, and Kepler 426-C is shown as a solid line. The orbital transit period of each planet's orbit may be found from the time elapsed between dips.

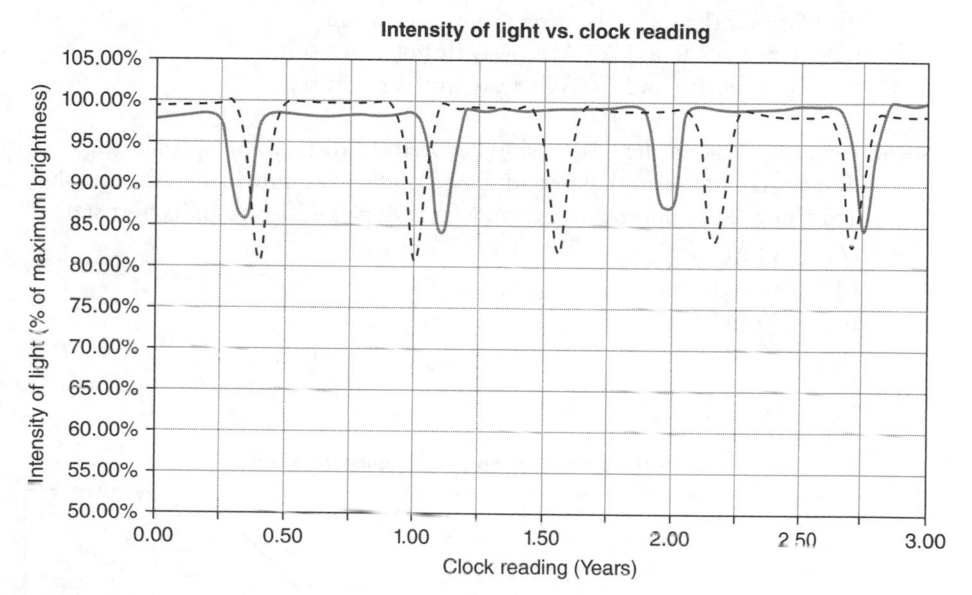

Figure 12.6

485. According to Figure 12.5, as the temperature of a star increases:

(A) the inner and outer radius increase at the same rate.

(B) the inner and outer radius increase, but the inner radius increases at a greater rate.

(C) the inner and outer radius increase, but the outer radius increases at a greater rate.

(D) the inner and outer radius increase, but the habitable zone narrows.

486. What is the most important factor that affects a planet's surface temperature?

(A) The temperature of the host star.
(B) The distance at which a planet orbits the host star.
(C) The transit period of a planet moving in front of its host star.
(D) The radius of the host star.

487. The passage states that Kepler 426-B orbits a star with a surface temperature of 6250 K. To support life, this planet must be a distance:

(A) Less than 1 AU away from its host sun.
(B) Greater than 2.2 AUs away from its host sun.
(C) Between 1.0 and 2.2 AUs away from its host sun.
(D) Between 1.5 and 3.6 AUs away from its host star.

488. Figure 12.7 shows the relationship between the distance of a planet from its host star and its orbital period. Based on the data provided in this graph and Figure 12.6, approximately how far is Kepler 426-B from its host star?

(A) 0.35 AU
(B) 0.50 AU
(C) 0.70 AU
(D) 1.50 AU

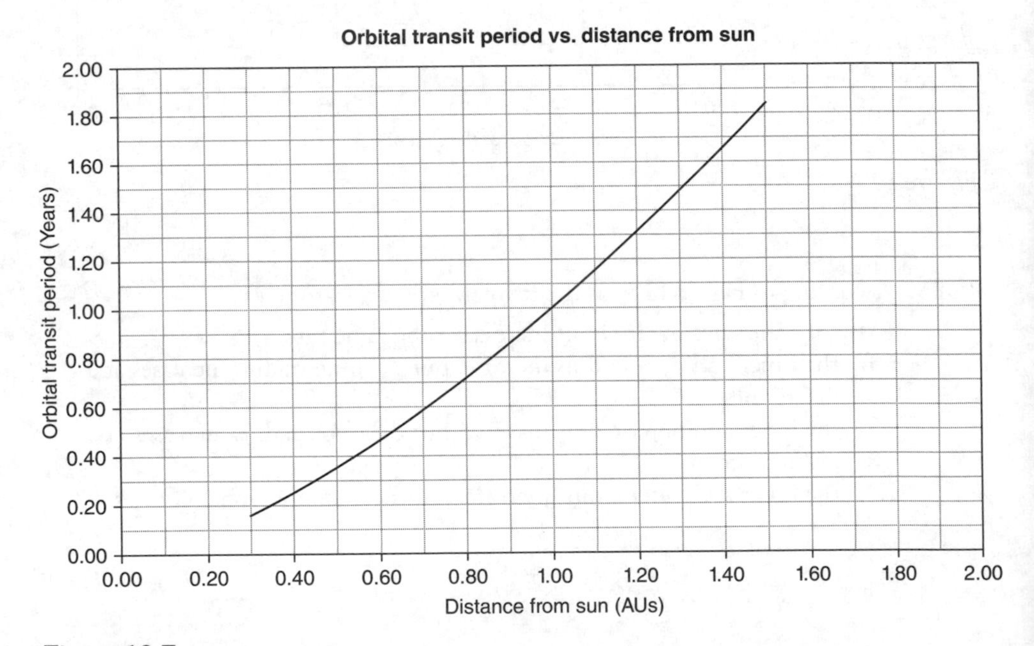

Figure 12.7

489. Two planets orbit a star with a temperature of 5500 K. Planet 1 has an orbital transit period of 0.60 years, while Planet 2 has a period of 1.8 years. Based on Figures 12.5 and 12.7, which of the following statements is true?

(A) Both planets are capable of supporting liquid water.
(B) Planet 1 is capable of supporting liquid water, but not Planet 2.
(C) Planet 2 is capable of supporting liquid water, but not Planet 1.
(D) Neither planet is capable of supporting liquid water.

Passage 45

We hear high or low pitches from musical instruments based on the frequency with which the instrument vibrates the air. Frequency is a measure of how many complete vibration cycles are produced each second, measured in units of Hertz. High pitches correspond with high frequency, whereas low pitches correspond with low frequencies. Some possible variables that may affect the frequency produced by a pipe are the length of the pipe, the diameter of the pipe, the ambient temperature of the environment, and whether the pipe is open at one end or both ends.

Three student groups are assigned the task of determining the relationship between various independent variables and the frequency of sound produced by the pipes. Group A has six pipes cut to different lengths, and the pipes are open at both ends. Group B also has six pipes cut to the same lengths as Group A; however, one of the ends of the pipes is closed. Table 12.6 shows the results from these two groups' investigations.

TABLE 12.6

Pipe #	Pipe length (m)	Frequency A (hertz)	Frequency B (hertz)
1	0.64	261	130
2	0.59	293	146
3	0.52	329	164
4	0.44	392	196
5	0.39	440	220
6	0.35	493	246

Group C is given six pipes with different diameters. Table 12.7 summarizes their findings.

TABLE 12.7

Pipe #	Pipe diameter (cm)	Frequency C (hertz)
1	3	196
2	5	220
3	7	195
4	9	192
5	11	193
6	13	194

490. Based on the data from Group A, estimate the length of pipe (open at both ends) that would play a musical note of F, which has a frequency of 349 cycles per second.
(A) 0.25 m
(B) 0.46 m
(C) 0.56 m
(D) 1.00 m

491. Which of the following is not a controlled variable in Group A's experiment?
(A) Ambient temperature
(B) Pipe diameter
(C) Pipe is open at both ends
(D) Frequency of note produced

492. Compared to a pipe open at both ends, what happens to the frequency produced by the same pipe when it is closed at one end?
(A) The frequency increases significantly.
(B) The frequency increases moderately.
(C) The frequency decreases significantly.
(D) The frequency remains approximately the same.

493. After the experimental data were gathered, Group C discovered that there was a problem with pipe 2. Which of the following is most likely the issue with that pipe?
(A) The pipe was too long.
(B) The pipe was too short.
(C) The pipe's diameter was measured incorrectly.
(D) The pipe was open at both ends.

494. Pipe 3 in Group C's experiment was also used in one of the other experiments. Which of the following best describes that pipe?

(A) 0.39 meters long and open at both ends
(B) 0.39 meters and closed at one end
(C) 0.44 meters and open at both ends
(D) 0.44 meters and closed at one end

495. Which of the following is the best summary of the findings from the three experiments?

(A) Shorter pipes closed at one end have the highest pitch.
(B) Long pipes open at both ends have the highest pitch.
(C) Large diameter and long pipes have the highest pitch.
(D) Shorter pipes open at both ends have the highest pitch.

Passage 46

When burned, cooking oils release different amounts of thermal energy into the environment. Chemists define "heat of combustion" as the amount of thermal energy released per unit mass of a substance after it combusts with oxygen at standard temperature and pressure. Figure 12.8 shows the apparatus for a basic combustion experiment used to calculate heat of combustion. The oil to be studied is placed inside the burner. When a wick is lit, the flame draws the oil up and allows it to react with oxygen in the air, and the thermometer measures the temperature of the water as it heats up.

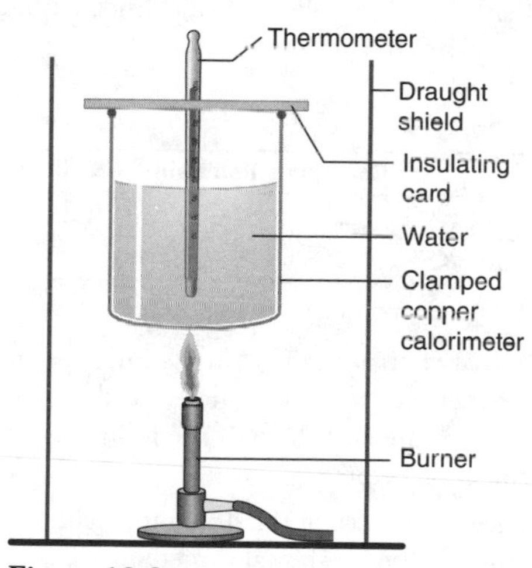

Figure 12.8

The thermal energy produced by the burning oil heats up the water in a copper cup, called the calorimeter. The following equation is used to calculate the thermal energy that is transferred from the burning oil to the water:

$$Q = m\,c\,\Delta T$$

Where:

Q = Thermal energy transferred to the water (measured in Joules)
m = mass of water (measured in grams)
c = Heat capacity of water (equal to 4.184 Joules/gram/°C)
ΔT = Temperature change of the water (measured in °C)

Cooking oils are made out of a variety of fatty acids. Fats can be classified as saturated or unsaturated based on the kind of bond between carbons. Saturated fats have all single bonds between carbons, while unsaturated fats have one or more double bonds. Table 12.8 shows the percentage of saturated vs. unsaturated fats in common cooking oils.

TABLE 12.8

Cooking Oil	Coconut	Red Palm	Safflower	Peanut	Olive
% Saturated	92	51	9	18	14
% Unsaturated	8	49	91	82	86

A student performs a combustion experiment by completely combusting 0.50 grams of the five different cooking oils in a calorimeter holding 100.0 grams of water. Table 12.9 summarizes the data.

TABLE 12.9

Type of Oil	Coconut	Red Palm	Safflower	Peanut	Olive
Initial water temperature (°C)	21	23	24	23	24
Final water temperature (°C)	26	27	25	25	26

496. A significant source of error in the combustion experiment described is that some of the thermal energy transfers into the environment instead of the water in the calorimeter. Which of the following would have the *least* effect on this source of error?

(A) The presence or absence of the draught shield.
(B) The amount of water in the calorimeter.
(C) The presence or absence of the insulating card.
(D) The proximity of the flame to the calorimeter.

497. What would you expect to happen to the temperature change in the experiment as well as the heat of combustion if twice as much oil was burned?

(A) Both the temperature change and the heat of combustion would be greater.
(B) Both the temperature change and the heat of combustion would be less.
(C) The temperature change would be less but the heat of combustion would remain the same.
(D) The temperature change would be greater but the heat of combustion would remain the same.

498. Which cooking oil has the largest temperature change during the experiment?

(A) Coconut
(B) Olive
(C) Safflower
(D) Red Palm

499. Assuming the same amount of oil is burned, how does the amount of energy transferred depend on the percent of saturation?

(A) As the percent of saturation increases, the energy transferred decreases.
(B) As the percent of saturation increases, the energy transferred also increases.
(C) The amount of energy transferred remains the same with percent of saturation.
(D) The energy transferred increases with saturation and then decreases.

500. Which calculation is correct for the heat of combustion (in Joules/gram) of the coconut oil using the information in the passage along with Table 12.9?

(A) $100.0 * 4.184 * 26$
(B) $100.0 * 4.184 * 5$
(C) $(100.0 * 4.184 * 26) \div 0.50$
(D) $(100.0 * 4.184 * 5) \div 0.50$

ANSWERS

Chapter 1: Test 1

1. (D) The endosperm may be yellow or white. Since the endosperm is located underneath the aleurone layer, its color is only visible when the aluerone is colorless. A purple or red aluerone masks the endosperm.

2. (A) The pericarp is the outermost layer of the corn seed, as shown in Figure 1.1. Since no color phenotypes related to the pericarp are mentioned, it can be inferred that the pericarp is colorless, allowing the color phenotype of the aleurone and endosperm to be visible.

3. (C) Table 1.1 lists three alleles (*C′*, *C*, and *c*) for the aleurone color inhibitor trait. The other traits listed each have two alleles (*Y* or *y*, *R* or *r*, and *P* or *p*).

4. (B) The aleurone may have a purple, red, or colorless phenotype. Colorless is not itself a kernel color phenotype because a colorless aleurone allows the endosperm color, either yellow or white, to be visible. Thus, there are four possible unique kernel color phenotypes: purple, red, yellow, and white.

5. (C) The aleurone color modifier genotype *pp* changes a purple kernel to red. A purple kernel results from a genotype that includes at least one *R* and one *C*. *RRCCppyy* is the only answer choice that satisfies these conditions. The endosperm color genotype (*yy*) does not matter in this case because the aleurone color masks it.

6. (B) The notation $P > p$ in the Allele Relationships column of Table 1.1 indicates that the allele *P* is dominant to the allele *p*, and thus the phenotype of *p* will be masked when *P* is present. This is supported by the fact that the genotypes *PP* and *Pp* both produce the same phenotypic outcome, while *pp* produces a different phenotypic outcome.

7. (D) Each kernel is a separate seed, or offspring, produced from the same pair of parents. Each seed gets the same genes, but the two specific alleles of each gene that each seed receives can be different.

8. (A) According to Table 1.1, a white phenotype can only result from a genotype containing the endosperm color combination *yy*. *rrCCppyy* is the only genotype listed that corresponds to a white phenotype.

9. (B) To have a yellow color, a kernel must have a colorless aleurone. Since both parents are red (and based on the relationships information in Table 1.1), the only genotype information known for certain about the kernels in Cross 2 is that each must have received the aleurone color modifier alleles *pp* from the parents. It is possible that the offspring kernels could receive from the parents some combination of *R* and *C* alleles that would produce a colorless aleurone, allowing the yellow endosperm to be visible.

10. (A) Cross 1 produced 100% yellow kernels. Repeating this cross will provide the greatest chance of producing more yellow kernels. Since Crosses 2 and 3 both produced lower percentages of yellow kernels, repeating either or both crosses would lower the overall probability of producing yellow kernels.

11. (C) Because yellow and white are two different alleles of the same gene (endosperm color), substituting a white parent for the yellow parent in Cross 3 would produce an ear with white kernels instead of yellow. Since endosperm color is masked by aleurone color, the kernels exhibiting purple aleurone color would not be affected by a change in endosperm color.

12. (C) The passage presents corn seed color as a trait that is influenced by four different genes whose influences are listed in Table 1.1. The passage does not mention an ability to change phenotype over time or any other functions of genes besides the influence on seed color.

13. (D) The only statement that can be completely supported by the information in the passage is that corn with a yellow phenotype must have a genotype that produces a colorless aleurone. Excluding mutations, all corn should have the same number and types of genes, though an exact genotype match to the yellow kernels in Cross 2 cannot be supported with the information available.

14. (C) The amount of friction between the airplane's surface and the air molecules determines the amount of drag, or backward force, on the plane. Like any other type of friction, drag opposes the motion of the airplane.

15. (D) Forces that oppose each other work on the same object but in opposite directions. Drag and thrust both act parallel to the motion of the airplane but in opposite directions. Lift and gravity also oppose each other.

16. (B) According to the passage, students tested three airplanes in Experiment 1 and four airplanes in Experiment 3. The type of paper used, number of students involved, and number of trials performed were constants in all three experiments.

17. (B) Based on the data in Table 1.3, no significant difference in horizontal distance can be identified among the three airplanes tested. In each trial, flat wings were the control group. Curving the wings upward appeared to cause a slight increase in distance in Trial 1, but a decrease in Trials 2 and 3. Curving the wings downward appeared to cause a slight decrease in distance in Trials 2 and 3, but no change in Trial 3.

18. (C) Flat wingtips represents the control group in Experiment 2 because this was the design that was unaltered from the original. Bending the wingtips up and down created two experimental groups to be compared to the flat-winged control group.

19. (D) Based on the data in Table 1.5, the paperclips placed midwing produced results closest to the results of the control group (no paperclips). This means that placing paperclips midwing had the least effect on the airplane's horizontal distance.

20. (C) Flat wings (Experiment 1), flat wingtips (Experiment 2), and no paperclips (Experiment 3) represent the unaltered airplanes. The horizontal distance traveled by these three planes ranged from 9.5 m (Table 1.3) to 11.1 m (Table 1.4). The approximate average horizontal distance of 10.5 m falls within the data range for the unaltered airplanes. The calculated average horizontal distance is 10.62 m.

21. (A) As shown in Table 1.4, curving the wingtips slightly upward increased the horizontal distance by approximately 2 m when compared to flat wingtips. This is the greatest positive effect of any variable in all three experiments.

22. (A) The student graph displays data for three airplanes. The line representing each airplane maintains a positive slope, indicating that the horizontal distance for all three airplanes increased with each subsequent trial. This corresponds to the data displayed in Table 1.3.

23. (D) In Experiment 3, adding paperclips to each paper airplane increased the plane's total weight. Adding weight to different locations along the airplanes altered their center of gravity by causing the location of the average weight to shift toward the paperclips.

24. (D) As shown in Table 1.4, bending the wingtips slightly downward decreased the horizontal distance by approximately 7 m when compared to flat wingtips. This is the greatest change in horizontal distance caused by any variable in all three experiments.

25. (C) In Experiment 1, the increase in horizontal distance for all three airplanes with each subsequent trial can most appropriately be attributed to the student throwing each airplane with an increasingly greater initial force. The force exerted by the student contributes to the amount of thrust on the airplane.

26. (B) In Table 1.4, wingtips bent up produced the greatest horizontal distance. In Table 1.5, no paperclips produced the greatest horizontal distance. No identifiable trend is present in the distance data in Table 1.3. Therefore, the combination of flat wings (Table 1.3), wingtips bent up (Table 1.4), and no paperclips (Table 1.5) can be expected to produce the airplane with the greatest horizontal distance.

27. (A) As illustrated in Figure 1.3, gram-positive bacteria have a cell wall composed of a thick layer of peptidoglycan and no outer membrane surrounding it. Gram-negative bacteria have a much thinner layer of peptidoglycan, surrounded by an outer membrane consisting of a lipid bilayer.

28. (C) As illustrated in Figure 1.3, both gram-positive and gram-negative cells have a periplasmic space located between the peptidoglycan layer and the cell membrane below.

29. (B) The lipid bilayer that makes up the outer membrane of gram-negative cells consists of an outer lipopolysaccharide layer and an inner phospholipid layer. Porins are transport channels that exist within this lipid bilayer, but they do not extend into the peptidoglycan layer. The thick peptidoglycan layer of gram-positive cells has no porins. This suggests that peptidoglycan is more permeable than the lipopolysaccharide-phospholipid bilayer, because the peptidoglycan layer does not contain any special transport channels, and both cell types must have the ability to transport substances into and out of the cell.

30. (D) As explained in the passage, ethyl alcohol is used as a decolorizer, not as a stain. Crystal violet is the primary stain, and safranine and fuchsin are two common counterstains, used in Gram staining.

31. (D) At the end of the Gram staining technique, cells appear either purple or red under the microscope. Gram-positive cells appear purple as a result of retaining the crystal violet stain. Gram-negative cells appear red because they do not retain the crystal violet and are subsequently stained by the counterstain.

32. (C) The purpose of adding iodine in the Gram staining technique is to form a complex with crystal violet that will become trapped inside gram-positive cells during the decolorization step. Application of the primary stain, crystal violet, must occur before iodine is added to allow this complex to form.

33. (D) Since gram-negative cells appear colorless after the decolorization step washes the primary stain out, the addition of a counterstain allows these cells to be seen more easily under a microscope. As noted in the passage, the red counterstain is lighter colored than the purple primary stain, so the addition of the counterstain does not affect the appearance of gram-positive cells.

34. (A) Ethyl alcohol acts as a decolorizer in the Gram staining technique. As described in the passage, the decolorization step degrades the outer membrane of gram-negative bacteria. Since this outer membrane consists of a lipid bilayer, it is most reasonable to assume that a chain of lipids will degrade in the presence of ethyl alcohol.

35. (A) Step 1a directs the user to continue to step 2 for gram-positive cells. Step 1b directs the user to continue to step 3 for gram-negative cells. Since the descriptions used in both steps 2 and 3 are identical, it can be assumed that rod-shaped and sphere-shaped bacteria species exist in both groups.

36. (B) Two of the genera (*Staphylococcus* and *Streptococcus*) in Table 1.6 are identified as gram-positive. Gram-positive bacteria have a thick peptidoglycan layer.

37. (A) By following Table 1.6 beginning at step 1, it can be determined that the genus *Pseudomonas* consists of bacteria that are gram-negative (step 1b), rod-shaped (step 3a), and do not ferment lactose (step 5b).

38. (D) By working backward through Table 1.6, it can be determined that both *Pseudomonas* and *Enterobacter* are gram-negative (step 1b) and rod-shaped (step 3a). *Pseudomonas* is identified as not fermenting lactose in step 5b. *Enterobacter* is identified as fermenting lactose (step 5a) and using citric acid (step 6a).

39. (B) Based on Table 1.6, the genus *Escherichia* has gram-negative cells, which should appear red after Gram staining because they absorb the counterstain instead of retaining the primary stain. This means that colorless gram-negative cells most likely indicate an error in the counterstaining step.

40. (A) According to the information in Table 1.6, *Staphylococcus* and *Streptococcus* are both gram-positive bacteria. Since gram-positive bacteria are less resistant than gram-negative bacteria to antibiotic treatment, it would be most effective to treat infections caused by bacteria in these genera with penicillin.

Chapter 2: Test 2

41. (C) According to the information in Table 2.1, European honeybee and jack jumper ant males are both haploid. The male swamp wallaby has one less chromosome than the female. Slime mold has no notation indicating a variability in chromosome numbers among individuals of the species.

42. (D) The coyote ($2n = 78$ chromosomes) and dog ($2n = 78$) both belong to the genus *Canis*. The horse ($2n = 64$) and the donkey ($2n = 62$) both belong to the genus *Equus*. The Bengal fox ($2n = 60$ chromosomes) and red fox ($2n = 34$) both belong to the genus *Vulpes*. These pairs of species demonstrate that two members of the same genus may have the same chromosome count (coyote and dog) or very different chromosome counts (Bengal fox and red fox).

43. (B) As the diploid number of chromosomes decreases, the complexity of the organism does not decrease, as exemplified by a human having fewer chromosomes than a potato. Neither does the complexity of the organism increase, as exemplified by oats having fewer chromosomes than a human. Therefore, based on the information in Table 2.1, no significant correlation can be identified between the complexity of organisms and their diploid chromosome count.

44. (B) The adder's-tongue fern has the greatest diploid number of chromosomes ($2n = 1{,}260$) in Table 2.1. A fern is a plant, which belongs to the kingdom Plantae.

45. (A) The dog has 78 diploid chromosomes, while a human has 46. The silkworm, *Bombyx mori*, falls between these numbers with a total of 54.

46. (B) The zebra fish and pineapple both have 50 diploid chromosomes. This means that 51 total chromosomes are contained in the nucleus of each somatic cell of both organisms.

47. (C) Gametes contain the haploid (n) number of chromosomes. Table 2.1 identifies the diploid ($2n$) number of chromosomes that are present in an organism's somatic cells. The horse (*Equus ferus caballus*) has a diploid number of $2n = 64$. This means the haploid number in gametes would be $n = 32$.

48. (C) Table 2.1 indicates that male European honeybees are haploid rather than diploid. This means a male has n chromosomes, while a female has $2n$ chromosomes. Since the diploid number of chromosomes is 32, $n = 16$. Therefore, females have 16 more total chromosomes than males.

49. (D) The total number of chromosomes in a somatic cell is referred to as the diploid number. The diploid number is represented by the term $2n$, in which n represents the number of chromosomes in a gamete from the same organism. The number of chromosomes in a somatic cell is twice the number of chromosomes in a gamete.

50. (C) As stated in the passage, the gametes of most species contain one complete set of chromosomes, and the somatic cells contain two complete sets. In polyploid species, the somatic cells contain more than two sets of chromosomes, but the gametes still contain half of the total number of chromosomes in a somatic cell. This means a gamete will contain more than one complete set of chromosomes.

51. (A) Table 2.1 indicates that alfalfa is a tetraploid species. The prefix *tetra* means "four." Therefore, as described in the passage, the somatic cells of a tetraploid species have four complete sets of chromosomes.

52. (B) The monoploid number (x) identifies the number of chromosomes an organism has in one set. For the hexaploid oat species *Avena sativa*, $6x = 42$. This means $x = 7$, or the number of chromosomes present in one set is 7.

53. (A) Table 2.1 indicates that the potato (*Solanum tuberosum*) is a tetraploid. Tetraploid organisms have four complete sets of chromosomes. As explained in the passage, the diploid number ($2n$) is always twice the haploid number (n). $2n$ always represents the total number of chromosomes in a somatic cell, regardless of whether the organism exhibits any type of polyploidy. Since a tetraploid has four complete sets of chromosomes, the total number of chromosomes ($2n$) must be divided by 4 to determine the monoploid number (x).

54. (B) According to Scientist 1, increased temperatures will help some crops and harm others. The growth rate for many crops increases with temperature, which is beneficial to those crops and the agriculture industry. However, increased temperature suppresses the reproductive ability of some crops, which endangers their survival.

55. (D) Based on the opinion of Scientist 1, a crop has its greatest yield close to its optimal growth temperature. Since average temperatures are predicted to rise as a result of climate change, plants with higher optimal growth temperatures are more likely to produce higher yields than those with lower optimal growth temperatures because they can better withstand the heat.

56. (D) Scientist 2 discussed the potential for crop destruction due to an increase in the frequency of flooding. Scientist 1 did not discuss the effects of flooding or any other extreme weather events.

57. (C) Scientist 2 discussed an increase in the production of weeds, pests, and fungi in response to the effects of climate change. Only Scientist 1 addressed the potential effect of climate change on seed production.

58. (A) In Scientist 1's opinion, increases in average temperature and carbon dioxide levels will have positive effects on some types of crops and negative effects on others. Scientist 2 only discusses the negative effects of these increases.

59. (B) Scientist 2 states that increased carbon dioxide levels will allow weeds, pests, and fungi to thrive. The increased presence of these organisms is predicted to impact crop yields negatively by competing with and damaging crops.

60. (B) In Scientist 2's opinion, climate change will lead to an increase in agricultural pests. An observed increase in pesticide use would provide evidence to support this opinion.

61. (D) Both scientists agree that increasing average temperatures will be beneficial to some organisms, though they do not agree on the particular organisms that will benefit. Scientist 1 suggests that increasing average temperatures will increase the growth rate of many crops, whereas Scientist 2 suggests that they will increase the growth of weeds, pests, and fungi, which will inhibit the growth of crops.

62. (D) Scientist 1 states that increased carbon dioxide levels will increase crop yields. It can be inferred that this change would be due to an increase in the rate at which crops carry out photosynthesis, since increased food production would lead to increased growth.

63. (A) Scientist 2 discusses the northward spread of weeds and pests as being detrimental to northern crops. While pests will directly damage crops, weeds will compete with crops for resources such as soil nutrients and water.

64. (D) According to Scientist 1, higher average temperatures and carbon dioxide levels will have direct, positive effects on many crops. This hypothesis can best be tested by gathering data on these two variables and the crop yields that result.

65. (A) According to Scientist 2, increased temperatures and carbon dioxide levels will cause the ranges of more heat-tolerant southern species to expand northward. As southern species expand northward, the ranges of northern species will likely narrow due to the increase in competition and new pests.

66. (B) Scientist 1 states that an increase in carbon dioxide causes crop yields to increase. However, once the optimal growth temperature for a crop is surpassed, crop yields will decline. Graph B best matches the scientist's description, with the maximum representing the point at which the optimal growth temperature has been reached.

67. (B) According to the photosynthesis equation, plants take in carbon from the environment in the form of carbon dioxide. This atmospheric carbon is used to produce organic carbon in the form of carbohydrates like glucose.

68. (D) According to the photosynthesis equation, plants use sunlight to convert carbon dioxide and water into a usable form of carbon (glucose). The by-product of this process is oxygen, which is released back into the environment.

69. (B) The passage states that each leaf was initially exposed to a light intensity of 300 μE/m²/s for a period of time to stimulate photosynthesis. According to Figure 2.5, this initial 300 μE/m²/s exposure began at approximately 5 minutes and ended at approximately 25 minutes, for a total duration of 20 minutes.

70. (A) As shown in Figure 2.7, the air temperature increased continuously over the course of the study. This continuous increase can best be attributed to the increase in light intensity over the same time. It can be inferred that increasing the intensity output of a light source also increases the heat output.

71. **(A)** The reactants of the photosynthesis equation are carbon dioxide and water. Figure 2.6 depicts the change in carbon dioxide concentration as recorded by sensors within the chamber over the course of the study.

72. **(C)** According to the passage, students manipulated the light intensity within the chamber and observed the effects on carbon dioxide concentration. This means that light intensity, depicted in Figure 2.5, is the independent variable.

73. **(D)** The lower the concentration measured within the chamber, the faster the leaf is absorbing carbon dioxide. This means that the lowest point on the graph in Figure 2.6, occurring at approximately 50 minutes, represents the highest absorption rate. According to Figure 2.5, the leaf is being exposed to a light intensity of 1,000 $\mu E/m^2/s$ at 50 minutes.

74. **(B)** In Table 2.2, the mean carbon dioxide exchange rate was the highest for the sunflower plant as compare to the other three plant types. This means that sunflower leaves absorb carbon dioxide fastest. Since carbon dioxide and oxygen are exchanged in a 1:1 ratio, it can also be stated that sunflower leaves release oxygen fastest.

75. **(C)** According to Table 2.2, rhoeo leaves showed the smallest standard deviation (1.3) across 10 trials of all four plant species studied. A low standard deviation indicates a low amount of variability across trials.

76. **(C)** The passage states that carbon dioxide and oxygen are exchanged in a 1:1 ratio during photosynthesis. As the carbon dioxide concentration in the chamber decreases, the oxygen concentration increases in direct proportion. Therefore, the oxygen concentration graph, just like the light intensity graph in Figure 2.5, would be the inverse of the carbon dioxide concentration graph in Figure 2.6.

77. **(B)** In Table 2.2, the mean carbon dioxide exchange rate for pothos leaves is 6.0 $\mu mol/m^2/s$. The means for both sunflower leaves (17.4 $\mu mol/m^2/s$) and water hyacinth leaves (14.3 $\mu mol/m^2/s$) are greater. A greater carbon dioxide exchange rate indicates a greater rate of photosynthesis.

78. **(C)** Figures 2.5 and 2.6 indicate that light intensity and carbon dioxide concentration within the chamber show an inverse correlation. This means that light intensity directly correlates to carbon dioxide absorption and therefore photosynthesis rate.

79. **(A)** In Table 2.2, the highest carbon dioxide exchange rate was recorded for the sunflower leaf (27 $\mu mol/m^2/s$) during Trial 3. A higher value indicates a faster exchange rate.

80. **(D)** Figures 2.5 and 2.6 indicate carbon dioxide concentration within the chamber is lowest when light intensity is highest. This means that the maximum carbon dioxide exchange rates identified in Table 2.2 were most likely recorded at 50 minutes, when light intensity was at its maximum (1,000 $\mu E/m^2/s$) and carbon dioxide concentration was at its minimum (approximately 880 ppm).

Chapter 3: Test 3

81. (D) Four possible ABO blood types (A, B, AB, and O) are listed in Table 3.1, and two possible Rh blood types (Rh+ or Rh–) are listed in Table 3.2. Each of the ABO blood types may be paired with either of the Rh blood types, resulting in eight possible medical blood types.

82. (B) As seen in Table 3.1, each ABO blood type is named for the antigen(s) present on the red blood cells. Blood type A has A antigens, and so on. Type O is so named because it has no antigens.

83. (A) As seen in Table 3.2, blood identified as Rh+ contains the Rh antigen but not the Rh antibodies.

84. (B) Blood type is identified by the antigens present on the red blood cells. Since only B antigens are present, the blood type would be B–. The presence of anti-A and anti-Rh antibodies means that A and Rh antigens are absent.

85. (C) Figure 3.1 indicates that 9% of the general population has type B+ blood and 2% has type B– blood, for a total of 11% of the general population with type B blood.

86. (D) According to Figure 3.1, only 1% of the general population has the AB– blood type.

87. (A) A higher percentage of individuals of Asian ethnicity have B+ blood (25%) than any other ethnicity. It can therefore be inferred that this blood type is more common in Asia than in the other three continents.

88. (C) A antigens are present on the red blood cells of both type A and type AB blood. The sum of type A+ (33%), type A– (6%), type AB+ (4%), and type AB– (1%) is 44%.

89. (B) In Table 3.3, 53% of Hispanic individuals, or a little over half, have O+ blood. The percentage of the general population that is Hispanic and has this blood type cannot be determined from the data provided.

90. (A) Based on the data in Table 3.3, an individual of African American ethnicity has an 18% chance of having B+ blood. This is greater than the 9% of the general population with the same blood type.

91. (C) As illustrated in Table 3.1, AB blood contains A and B antigens but no antibodies. Since no antibodies are present, the antigens on the donated blood will not be attacked, leading to no immune reaction to the donated type A blood.

92. (B) As illustrated in Tables 3.1 and 3.2, blood type O– has no antigens on its red blood cells. If no antigens are present, there is nothing to trigger antibodies to attack.

93. (C) An individual can safely receive a transfusion of his or her own blood type because the donor blood contains the exact same antigens (A antigens in this case) as are already present, triggering no antibody attack. Any individual can also receive a transfusion of O– blood because it is the universal donor, containing no antigens at all.

94. (C) According to Table 3.3, 1% of the Caucasian population has AB– blood. Figure 3.1 indicates that 1% of the general population also has this blood type.

95. (D) In Figure 3.2, the cathode is shown at the top of the diagram and the anode is at the bottom. When the power supply is turned on, the resulting electric current causes the DNA samples loaded in the wells to travel away from the negative charge produced by the cathode and toward the positive charge of the anode.

96. (B) A DNA ladder is a solution containing DNA fragments of known sizes. When an electrophoresis procedure is run, the migration of these fragments provides a reference by which to estimate the sizes of DNA fragments in the samples.

97. (A) Since Table 3.4 identifies the voltage range as 0.25–7 V/cm, a voltage of 6 V/cm is relatively high. Table 3.4 indicates that a high voltage may cause smearing or poor resolution of large DNA fragments.

98. (C) In Table 3.5, a 0.5% agarose concentration is recommended to resolve DNA fragments of 1–30 kb, or a 29 kb range. This range decreases incrementally as agarose concentration increases. A 1.5% concentration is recommended to resolve DNA fragments from 0.2–0.5 kb, which is only a 0.3 kb range.

99. (C) Based on the information in Table 3.5, DNA fragments of 0.5–0.7 kb are relatively small. Table 3.4 indicates that a high agarose concentration provides a sharper resolution of small DNA fragments, and low voltage may cause fragments of less than 1 kb to diffuse in the gel. This means that a high agarose concentration (1.2%) and a high voltage (5 V/cm) would provide the best results for DNA fragments of this size. Table 3.5 indicates that the highest agarose concentration (1.5%) is only appropriate for fragments smaller than 0.5 kb.

100. (A) According to the passage, large DNA fragments travel more slowly than small fragments. This means the largest DNA fragments will appear closest to the cathode, because they will have traveled the least distance from the wells.

101. (B) The passage indicates that Allele 2 is larger than both Alleles 1 and 3. Therefore, Allele 2 will travel through the agarose gel more slowly than the other two.

102. (D) According to the passage, a single band that appears darker is an indicator of two copies of the same allele. Individuals E and K both appear to have two copies of Allele 2, while N and O appear to have two copies of Allele 1.

103. (A) Based on information in the passage, Allele 2 is the largest of the three alleles. Therefore, Allele 2 is indicated by the band that has traveled the least distance down the gel. The size of this band is approximately 3.0 kb.

104. (B) The combination of Alleles 1 and 2 occurs most frequently in Figure 3.3, appearing in a total of six lanes (Lanes A, B, D, I, L, and M).

105. **(D)** In Figure 3.3, all three alleles appear to be present in the sample in Lane J. The most reasonable explanation is that this lane contains DNA from more than one individual, since each individual can carry only two copies of one gene.

106. **(C)** In Figure 3.3, all allele combinations (Alleles 1 and 1, 1 and 2, 1 and 3, etc.) are represented in at least one lane except the combination of two copies of Allele 3.

107. **(C)** The three alleles in Figure 3.3 appear to be approximately 2–5 kb in size. The most appropriate agarose concentration for this size range is 1.0%. A higher concentration of agarose is more appropriate for DNA fragments that are smaller than these, and a lower concentration is more appropriate for larger DNA fragments.

108. **(D)** According to Table 3.4, low agarose concentration and low voltage both cause longer run times (possibly over multiple days). Therefore, DNA fragments should migrate most slowly when agarose concentration and voltage are both at their minimum. This corresponds to an agarose concentration of 0.5% and a voltage of 0.25 V/cm.

109. **(B)** A *stimulus* is a change in the environment, such as the sounding of a tone or the appearance of an image. A *response* is the individual's reaction to the stimulus. The experiment descriptions identify the response time as the time between the sounds of a tone or the appearance of an image (stimulus) and the student pressing the spacebar (response).

110. **(A)** In Experiment 1, the student pressed the spacebar in response to the sounding of a tone, so the sounding of the tone was the stimulus.

111. **(D)** Both Experiments 2 and 3 tested response time to auditory and visual stimuli. In Experiment 2, three successive trials were performed using an auditory stimulus, and then three were performed using a visual stimulus. In Experiment 3, the type of stimulus alternated with each new trial. All other conditions were held constant between the two experiments.

112. **(D)** According to Table 3.7, the response time to an auditory stimulus is consistently lower (faster) than the response time to a visual stimulus. A faster response time indicates that auditory processing occurs faster than visual processing, suggesting that the sense of hearing is more acute.

113. **(D)** All three experiments used a stimulus duration of 400 ms for at least some trials. In Experiment 1, trials 4 through 6 used a 400 ms tone length. A 400 ms stimulus duration was used on all trials during Experiments 2 and 3.

114. **(B)** The fastest reaction time was 142 ms. This occurred during Trial 5 of Experiment 1 as well as Trials 2 and 3 of Experiment 2. All three of these trials recorded response times to an auditory stimulus lasting a duration of 400 ms.

115. **(A)** In Table 3.6, Trials 1 through 3 were conducted using a 200 ms tone length and Trials 4 through 6 used a 400 ms tone length. The average response times were approximately 10 ms faster for the trials using the longer tone.

116. (C) In Tables 3.7 and 3.8, the response times for the auditory stimulus were consistently more than 40 ms faster than for the visual stimulus. It can be inferred that a faster response time is the result of a signal reaching the brain faster.

117. (A) The response times recorded for auditory stimuli across all three experiments range from 142–158 ms. It is therefore reasonable to infer that the typical auditory response time range is 140–160 ms.

118. (C) According to the description of Experiment 2, 6 total trials were conducted, and each trial consisted of 10 stimulus-response cycles. This means that a total of 60 responses to a stimulus were recorded during Experiment 2.

119. (C) Both auditory and visual response times increased over the course of Experiment 3. Therefore, the graph for the data in Table 3.8 should consist of two lines, each with a small positive slope. Response time to the visual stimulus was consistently higher than to the auditory stimulus, so the visual line should be above the auditory line.

120. (B) The student's new experimental design should be the same as in Experiment 1, but using a visual stimulus (image) instead of an auditory one (tone). The student can then test the difference in response time to an image shown for 200 ms versus an image shown for 400 ms.

121. (D) The slowest auditory response time occurred during Trial 1 of Experiment 1. The response time was 158 ms.

122. (C) In Tables 3.6 and 3.7, the response time for a particular stimulus decreased (got faster) over subsequent trials. In Table 3.8, the response time for a particular stimulus increased (got slower) as the type of stimulus alternated with each trial. This suggests that repetitive exposure to the same stimulus improves an individual's reaction time.

Chapter 4: Test 4

123. (B) Organic molecules are found only in the bodies or products of living organisms. Carbohydrates, lipids, proteins, and nucleic acids are all produced by living organisms. Although living organisms are composed largely of water, water is inorganic because it is not produced by living organisms.

124. (D) Both the primordial soup theory and the hydrothermal vents theory assume that organic molecules can be produced by reactions that cause the rearrangement of atoms in certain inorganic molecules. Scientists believe this to be an important step toward the existence of life on the earth.

125. (A) In Figure 4.1, the heat source is located underneath the small sphere that simulates the water in the earth's oceans. According to the diagram, the ocean (small sphere) supplies water vapor for the reactions in the atmosphere (large sphere). The heat source facilitates the production of water vapor.

126. (A) According to the passage, ammonia was believed to be a major inorganic component of the primitive atmosphere that contributed to the production of organic molecules.

127. (B) In Figure 4.1, the small sphere simulates the ocean, which provides water vapor for the primitive atmosphere. The actual reaction that produces organic molecules occurs as the electrical current is passed through the large sphere that simulates the primitive atmosphere. The condenser then cools the gases in the atmosphere, allowing newly produced organic molecules to condense into solution and travel to the trap for sampling.

128. (D) The passage states that the minimum and maximum temperatures around hydrothermal vents are 4°C and 300°C, and organic molecules are only stable within a narrow temperature window. An optimal temperature range of 4°C to 25°C meets both of these criteria.

129. (C) Both theories identify water as playing an integral role in the development of organic compounds on Earth. The primordial soup theory identifies water as one of the four major components of the primitive atmosphere. The hydrothermal vents theory identifies deep ocean water as the site of organic compound formation.

130. (C) The two theories disagree on the energy source used to fuel the reactions that originally produced organic compounds. The primordial soup theory identifies atmospheric lightning as the energy source, whereas the hydrothermal vents theory argues that the energy came from within the earth.

131. (D) The primordial soup theory is based on the assumption that the primitive atmosphere was mainly composed of methane, ammonia, hydrogen, and water vapor. Helium is not believed to have made up a significant percentage of that atmosphere.

132. (C) The presence of a constant electrical charge is a potential limitation of the Miller-Urey apparatus, because it causes experimental conditions to differ from the conditions scientists believe actually existed in the primitive atmosphere. The constant charge simulates a constant supply of lightning that scientists do not believe existed.

133. (C) The hydrothermal vents theory identifies the location of organic molecule formation as the deep ocean. Hot gases and energy from inside the earth enter the deep ocean through hydrothermal vents. The theory suggests that organic molecules are then produced within the temperature gradient generated at hydrothermal vents.

134. (B) The Miller-Urey experiment supports the primordial soup theory because it successfully produced organic compounds under the same conditions as scientists believed existed on primitive Earth. The experiment was specifically designed to investigate this theory and does not provide direct evidence to support or reject the hydrothermal vents theory.

135. (A) Based on the passage, the constant release of hot gases into the cold deep ocean water produces a temperature gradient. This temperature gradient is believed to provide sufficient conditions for the production of organic molecules.

136. (B) The primordial soup theory assumes that the compositions of the primitive and current atmospheres vary widely. This assumption led to the development of the Miller-Urey experiment, which produced various organic molecules from the compounds thought to be most abundant in the primitive atmosphere. The current atmosphere is composed mostly of nitrogen and oxygen.

137. (D) The passage defines the normal boiling point as the temperature at which vapor pressure and standard atmospheric pressure are equal. The normal boiling point of a substance does not change.

138. (B) The boiling point of a compound is always the temperature at which the liquid's vapor pressure is equal to atmospheric pressure. A change in atmospheric pressure results in a similar change in the vapor pressure required to induce boiling.

139. (B) According to Figure 4.3, hexane will boil at 0°C when the vapor pressure is decreased to approximately 50 mmHg. The vapor pressures required to boil heptane and octane at 0°C are both less than 50 mmHg. The vapor pressure required to boil pentane at 0°C is approximately 200 mmHg.

140. (D) In Figure 4.3, the normal boiling point for pentane is about 36°C. Increasing the temperature to 40°C requires an increase in vapor pressure as well. Thus, 850 mmHg is the best estimation for this vapor pressure.

141. (A) The difference between normal boiling points for each consecutive pair of alkanes in Figure 4.3 is about 30°C. Adding 30°C to the boiling point of heptane (98.4°C) provides an approximate value of 128°C for octane's normal boiling point.

142. (A) In Figure 4.4, the alkanes are near the alkenes and alkynes. These three functional groups exhibit similar boiling point trends in the graphs, which happen to be the lowest of the functional groups included.

143. (C) According to Figure 4.4, a 2-carbon alcohol has an approximate boiling point of 75°C. Figure 4.4 indicates that a 4-carbon ketone would have approximately the same boiling point.

144. (C) The passage indicates that stronger bonds require higher temperatures to break. The alkanes, alkenes, and alkynes have the lowest boiling points in Figure 4.4. The lowest boiling points mean that the bonds (Van der Waals) in these functional groups are the easiest to break.

145. (A) According to its molecular formula, caproic acid contains six carbon atoms. Figure 4.4 illustrates that a 6-carbon carboxylic acid will have a boiling point of approximately 200°C.

146. (B) According to the passage, strong bonds require higher boiling points to break, while weak bonds require lower boiling points. Carboxylic acids have the highest boiling points (Figure 4.4) and therefore the strongest bonds (double hydrogen). Ordering the functional groups in Figure 4.4 from highest to lowest boiling points allows for the identification of bonds from strongest to weakest.

147. (D) In Table 4.2, the molecular weights of each molecule are similar, but the boiling points are not. This indicates that molecular weight does not directly affect boiling point.

148. (A) In Table 4.2, n-Butanol contains four carbon atoms and has a boiling point of 117°C. In Figure 4.4, a boiling point of 117°C most closely resembles the boiling point of a 4-carbon member of the alcohol group.

149. (D) Based on the data in Table 4.1 and Figure 4.4, double hydrogen bonds are the strongest, followed by single hydrogen bonds, dipole-dipole bonds, and Van der Waals bonds. At the same vapor pressure, stronger bonds require a higher temperature to break than weaker bonds. Since single hydrogen bonds are stronger than dipole-dipole bonds, single hydrogen bonds require a higher boiling point.

150. (B) In both Figures 4.3 and 4.4, the boiling point increases as the number of carbon atoms increases within a specific functional group. This means the number of carbon atoms can be used to predict the relative boiling point for compounds containing a known number of carbon atoms in a known functional group. A common number of carbon atoms among compounds from different groups does not necessarily indicate similarities in boiling point.

151. (A) A closed ecosystem is characterized by a lack of migration into and out of it. There are several possible causes, one of which is geographic isolation. The passage states that Shebay Park consists of a group of isolated islands. Since the islands are separated from the mainland, migration to and from this ecosystem is rare.

152. (B) In Figure 4.5, the peccary population is shown to feed on four different types of plants. A consumer that feeds exclusively on plant matter is termed an herbivore.

153. (C) In Figure 4.6, the peak in the jaguar population occurred around 1990. In this year, the population consisted of 50 individuals.

154. (A) In Figure 4.6, the smallest value for the peccary population was 500 individuals. This occurred in roughly 2006.

155. (A) According to Figure 4.6, the peccary population experienced many consecutive years of positive population growth beginning around 1990. Immediately preceding this growth period, the jaguar population was devastated by exposure to feline leukemia (1989). It can be inferred that the sharp decline in the jaguar population caused a decrease in predatory pressure on the peccary population, thereby allowing the latter population to increase.

156. (D) Consumers in an ecosystem are identified according to how far removed they are from the ecosystem's producers (plants). Primary consumers are herbivores that feed only on plants. Secondary consumers feed on primary consumers. According to the food web in Figure 4.5, two secondary consumers exist in the ecosystem—the fox and the jaguar.

157. (D) The food web in Figure 4.5 shows that nutria and duck share common food sources and a common predator with the peccary. These similarities indicate that the three populations occupy similar niches within the ecosystem.

158. (B) In Figure 4.6, the sharp declines in the jaguar (1990) and peccary (2004) populations coincide with the occurrence of rare environmental events. The passage indicates that the 1990 decline in the jaguar population can be attributed to the introduction of feline leukemia in 1989, and the 2004 decline in the peccary population can be attributed to severe winter conditions and a tick outbreak. Each of these events reduced the immediate ability of individuals to survive, resulting in a population size that was drastically reduced in a matter of two years.

159. (A) The passage states that an outbreak of ticks occurred in 2004. This coincides with the severe reduction in peccary population size beginning that year, suggesting that the parasite outbreak contributed to the decrease in the population.

160. (D) Several factors influencing the peccary population size are discussed in the passage, including interactions with other species and environmental conditions.

161. (B) While most species in the food web (Figure 4.5) are directly linked to the peccary, the squirrel shares no direct connection with that population. Squirrels and peccaries have no common predators or food sources, therefore the squirrel should be less affected by a change in the peccary population than other species in the food web.

Chapter 5: Test 5

162. (B) An invasive species is both non-native to an ecosystem and harmful in some way. Non-native species that move into a new ecosystem will disrupt that ecosystem by competing with native species for resources and introducing new feeding relationships into the food web.

163. (C) The passage indicates that water hyacinths are able to withstand fluctuations in pH, but it does not indicate that these plants induce changes in pH. Water hyacinths upset freshwater ecosystems in a number of ways, but not by altering pH.

164. (B) In Table 5.1, the final weed combination studied was the combination of all four weed species. This same combination is not present in Table 5.2.

165. (A) Plant density is identified in the description of Study 1 as the total number of water hyacinth plants within a sample area. A sample area was defined as 1 square meter.

166. (D) The sample areas in which water hyacinth was found growing alone provide the control group for Study 1. These sample areas indicated water hyacinth growth when there was no competition from other weed species.

167. (C) In Table 5.1, the weed combination of *E. crassipes* and *Commelina* sp. exhibits the least difference from the control group (*E. crassipes* alone) out of all possible combinations. This means that *Commelina* sp. exerts the least competitive pressure on water hyacinth.

168. (D) In Table 5.2, water hyacinths grown alone were shown to have an average height of 8.69 cm. When grown with *Commelina* sp. and *Justicia* sp. in a greenhouse, the average water hyacinth height was shown to increase to 8.80 cm and 8.88 cm, respectively. The increase in growth suggests a positive effect of these competitor species on water hyacinth height.

169. (A) The number of leaves per plant for a water hyacinth grown alone was higher in Table 5.2 than in Table 5.1. The values for fresh weight and plant height were both much lower in Table 5.2 than in Table 5.1. Total biomass was not calculated in Study 2.

170. (A) In the description of Study 1, total biomass was described as being calculated by multiplying plant density by fresh weight. Since plant density is a component of total biomass and was not recorded in Study 2, total biomass could not be calculated for this study.

171. (A) In both studies, *V. cupsidata* was shown to decrease water hyacinth growth by all parameters, except for fresh weight in Study 1. These data support the claim that *V. cupsidata* has the most negative impact on water hyacinth growth.

172. (B) In Table 5.2, all three competitor weeds were shown to reduce water hyacinth fresh weight, thus leading to lighter plants than when water hyacinth grew alone. For both plant height and leaves per plant, at least one competitor species was shown to increase water hyacinth growth.

173. (C) In Study 2, scientists removed young weed plants from the Kagera River and grew them in a greenhouse. Within the greenhouse, the researchers had more control over the environmental conditions to which the plants were exposed, thus controlling any environmental factors (other than competitor species) that may influence water hyacinth growth.

174. (D) In Table 5.1, the average fresh weight of water hyacinth growing in the presence of all three competitor species was 342 g. This value is approximately equal to the mean (average) water hyacinth fresh weight of 368 g when growing in the presence of each competitor species individually.

175. (D) Throughout both studies, *V. cupsidata* consistently showed the most negative effect on water hyacinth growth. This is true for all growth parameters except one (fresh weight in Study 1). *Commelina* sp. and *Justicia* sp. both showed mixed effects on water hyacinth growth, depending on the growth parameter observed and the study environment. Therefore, increasing the presence of *V. cupsidata* alone can be predicted to significantly reduce the water hyacinth population.

176. (D) The passage provides the formula for gravitational potential energy as $PE_g = m \times g \times h$, in which m represents the object's mass, g represents acceleration due to gravity, and h represents the object's height above the ground. Acceleration due to gravity does not change on Earth. Therefore, g is a constant.

177. (A) Based on the formula $PE_g = m \times g \times h$, increasing the value of h will cause a similar increase in the value of $PE_{(g)}$. This means the total gravitational potential energy of an object will double when the object's height is doubled.

178. (B) As stated in the passage, the amount of potential energy at the beginning (point A) and end (point C) of the roller coaster would be equal in a frictionless environment. This is explained by the law of conservation of energy, which states that energy cannot be lost or gained within a system. For the marble to have the same potential energy at points A and C, the roller coaster must have the same drop height and hill height.

179. (C) As stated in the description of Experiment 1, altering the drop height changed the marble's initial amount of gravitational potential energy because, according to the formula provided in the passage, an object's gravitational potential energy is dependent on the object's height.

180. (B) The maximum drop height used in both experiments was 1.2 m. This drop height was tested in Trial 4 of Experiment 1 and held constant for Experiment 2.

181. (C) In Table 5.3, drop height and hill height exhibit a direct relationship. As drop height increases, so does hill height. Figure 5.4 provides a visual representation of this direct relationship between the two variables.

182. (A) Table 5.4 indicates that when drop height was held constant, a longer horizontal distance required a shorter hill height. This means that although each marble began at the same height, marbles that traveled a longer track had more energy transformed to heat and sound through frictional dissipation. This increase in frictional dissipation left less kinetic energy available to propel the marble as far up the hill.

183. (D) The smallest hill height recorded in Experiment 2 was 0.97 m. This height was recorded during Trial 3 as the result of a drop height of 1.2 m and a horizontal distance of 1.5 m.

184. (D) In Experiment 1, drop height was the independent variable. Students varied the drop height to determine the effects on hill height. In Experiment 2, drop height was held constant to determine the effects of horizontal distance on hill height. Neither experiment measured drop height as the dependent variable.

185. (A) To study the relationship between marble mass and hill height, students should hold drop height constant, with marble mass becoming the independent variable and hill height remaining the dependent variable. To accommodate this change in the data table, an additional column for marble mass should be added, just as the horizontal distance column was added in Table 5.4.

186. (A) As stated in the description of Experiment 1, Point A indicates the starting height (or drop height) of the marble. Because of frictional dissipation, the marble has the greatest gravitational potential energy at Point A. Once the marble begins moving along the track, frictional dissipation begins transforming some of the marble's energy to heat and sound, leaving less energy available for the marble itself.

187. (B) Changes in a dependent variable are measured to determine the effects of the independent variable. In both experiments, hill height was the dependent variable. Changes in hill height provided information on the effects of the independent variables on initial gravitational potential energy (Experiment 1) and frictional dissipation (Experiment 2).

188. (C) Mechanical energy is the energy related to an object's motion and position, and it consists of the sum of an object's potential and kinetic energy. During the roller-coaster experiments, mechanical energy was transformed between potential and kinetic forms. Frictional dissipation also caused some mechanical energy to be transformed to thermal energy (heat) and sound. No transformation to chemical energy occurred.

189. (B) In Table 5.4, hill height decreased as horizontal distance increased. Therefore, increasing the longest horizontal distance in Trial 3 (1.5 m) to 1.75 m would further decrease the hill height.

190. (A) According to Table 5.5, all organisms in the phylogenetic tree are members of the class Mammalia. This means that all of the organisms are mammals.

191. (C) *Panthera pardus* is the scientific name for leopard. Table 5.5 identifies the leopard as belonging to the family Felidae.

192. (B) All organisms listed in Table 5.5 share a common kingdom, phylum, class, and order. The organisms diverge into different taxa beginning with the family level.

193. (D) Table 5.5 lists *Canis latrans* as the genus and species of the coyote.

194. (C) In a phylogenetic tree, a species is most closely related to the species with which it shares the most recent common ancestor, or node. According to Figure 5.6, the European otter (*Lutra lutra*) shares the most recent common ancestor with *Taxidea taxus*.

195. (B) In a phylogenetic tree, the number of common ancestors shared by two species is indicated by the number of nodes shared by those species. In Figure 5.6, *Panthera pardus* diverges from the other four species at the very first node, indicating that it shares only one common ancestor with the other four species.

196. (D) The passage states that the length of the horizontal lines on a phylogenetic tree indicates the relative divergence time between species. *Canis latrans* and *Canis lupus* are connected to their most recent common ancestor by the shortest lines, indicating that these two species diverged most recently.

197. (A) The passage identifies an extant species as one that is currently living. *Taxidea taxus*, listed along the right side of the phylogenetic tree in Figure 5.6, is currently living.

198. (B) The seven-level classification system used in Table 5.5 classifies organisms using a hierarchical system that goes from broadest grouping (kingdom) to most specific (species). As shown in Table 5.5, organisms that share a certain taxonomic level also share the same higher taxonomic levels, but they may diverge in lower levels. This means that species belonging to the same order must also belong to the same phylum and kingdom, but they may belong to a different order.

199. (C) The lynx belongs to the family Felidae. *Panthera pardus* also belongs to this family. Based on the phylogenetic tree in Figure 5.6, the lynx shares the most recent common ancestor with *Panthera pardus*.

200. (D) The lynx belongs to a different family than the gray wolf, but the same taxa for order and above. This means that the lynx and gray wolf share four common taxonomic levels.

201. (B) The five species in Table 5.5 belong to three different families, but they all belong to the same order (Carnivora). This means the most recent common ancestor shared by all five species was also a member of the Carnivora taxon.

202. (C) A clade must include all extant taxa that have descended from a particular ancestral taxon. In Figure 5.6, *Canis latrans, Canis lupus,* and *Lutra lutra* do not constitute a complete clade because their most recent common ancestor is also the ancestor of *Taxidea taxus.*

203. (D) According to Table 5.5, members of the family Mustelidae are also members of the order Carnivora. This is the same order to which the coyote belongs, since all species in Table 5.5 belong to this order. There is not enough information to determine to which member of Mustelidae in Table 5.5 the wolverine is more closely related.

Chapter 6: Test 6

204. (C) The code for building a protein is stored in the cell's DNA. As with any protein, the first step in building a tropomyosin protein is transcribing the α-TM gene into mRNA. After RNA splicing, this mRNA transcript can be used to build the protein.

205. (A) The maximum number of exons present in an mRNA transcript is 10. Both types of striated muscle exhibit mRNA transcripts with 10 exons.

206. (D) Exon 4 is present in all seven transcripts in Figure 6.1. This suggests that Exon 4 is a constitutive exon necessary for all tropomyosin proteins.

207. (B) The mRNA transcript found in the brain contains six constitutive exons and one alternatively spliced exon. Exon 7 is present in the brain but not in all tissue types.

208. (C) Exons 2 and 3 do not appear together in any of the seven mRNA transcripts in Figure 6.1. Exons 11 and 12 also do not appear together.

209. (D) Since two samples of striated muscle are shown in Figure 6.1, it can be assumed that one sample is skeletal muscle and the other is cardiac muscle. Examining both mRNA transcripts shows that the sample labeled Striated muscle ends with Exon 11, while the sample labeled Striated muscle′ ends with Exon 12. No other differences exist between the two types.

210. (B) According to Figure 6.2, the α-TM gene contains 12 total exons. Though the number of exons in the mRNA transcripts varies, the number of exons in the actual gene does not change.

211. (B) According to Table 6.1, Exon 4 codes for amino acids 81–125. This is a total of 44 amino acids, which is the greatest number of amino acids coded by any single exon in the table.

212. (D) According to Table 6.1, either Exon 10 or 12 codes for amino acids 258–284. The brain mRNA transcript contains neither Exon 10 nor 12.

213. (A) Table 6.1 shows that Exons 10 and 12 both code for amino acids 258–284. Though all other transcripts only contain one of the two exons, striated muscle′ contains both. This means that the striated muscle′ mRNA transcript contains two copies of the code for amino acids 258–284.

214. (B) Based on Figure 6.1, the difference between a myoblast and a smooth muscle transcript involves Exons 2 and 3. Exon 3 is present in the myoblast transcript but is absent from the smooth muscle transcript, where it is replaced by Exon 2.

215. (C) According to Table 6.1, Exon 11 codes for no amino acids. This suggests that Exon 11 must contain an untranslated region instead of an amino acid code.

216. (A) In Figure 6.1, the first four mRNA transcripts are from different types of muscle (or muscle precursor) cells. The last three transcripts are from different types of nonmuscle cells. All four of the muscle cell transcripts contain either Exon 2 or Exon 3. None of the nonmuscle transcripts contain either of these exons. Therefore, it can be inferred that the extra function of tropomyosins in muscle cells could be related to the sequence of amino acids (amino acids 39–80) coded by Exons 2 and 3.

217. (D) According to Figure 6.1, the hepatoma is the only tissue type from which Exon 7 is missing. This implies that a loss of Exon 7 may be correlated to tumor formation.

218. (C) The passage identifies neonicotinoids as a class of pesticides thought to be harmful to honeybees.

219. (B) Scientist 1 states that neonicotinoids have been shown to increase honeybees' susceptibility to disease and parasites. Though no direct link between neonicotinoids and honeybee population loss has been found, Scientist 1 believes neonicotinoids to be an indirect factor.

220. (A) Scientist 1 lists almonds, apples, onions, and carrots as crops that rely almost exclusively on honeybees for pollination. Cherries are not mentioned.

221. (B) Scientist 1 states that any measures with the potential to improve honeybee health, such as a neonicotinoid ban, should be attempted. Scientist 2 states that a neonicotinoid ban should not be attempted because neonicotinoids are not among the greatest threats to honeybee health, and efforts should instead be focused on the strongest known threats.

222. (A) According to Scientist 1, reproductive rates of honeybee-dependent crops vary directly with the availability of honeybees. This would also apply to honeybee-dependent producers in the natural ecosystem. Figure 6.3 demonstrates a direct relationship between honeybee population size and producer reproductive rate. As honeybee population size increases, so does the producer reproductive rate.

223. (A) Scientist 2 states that varroa mites and disease are the greatest known threats to honeybee health and suggests that efforts to improve that health should focus on these threats.

224. (D) Scientist 2 identifies water that is scarce or of poor quality as an environmental stressor to honeybees. Scientist 1 does not mention water quality as a factor affecting honeybee health.

225. (C) Scientist 1 states that honeybees are important to the overall ecosystem because plants at the base of the food web depend on pollination by honeybees. This indicates that Scientist 1 believes transferring pollen between plants is honeybees' most important contribution to natural ecosystems.

226. (D) Scientist 2 indicates that greater exposure to neonicotinoids does correlate slightly to a decrease in honeybee population size. The graph in option D demonstrates a weak negative correlation between pesticide exposure and honeybee population size. Though the data points do not form a tight line, a negative average slope is identifiable.

227. (C) The passage states that the European Union recently instituted a two-year ban on neonicotinoids in an effort to improve the health of European honeybee populations. A doubling of European honeybee populations over the next five years would indicate that the neonicotinoid ban was effective in improving honeybee health. This supports the opinion of Scientist 1, who believes a neonicotinoid ban would be effective in the United States.

228. (B) Both scientists discussed disease and parasites (varroa mites) as contributing to the decline of honeybee populations. Scientist 2 asserts that disease and parasites are the greatest threats to honeybee health. Scientist 1 asserts that any measures available to improve honeybee health should be taken. Therefore, it can be inferred that both scientists would support efforts to improve disease and parasite prevention.

229. (A) Scientist 1 states that 23% of crops in the United States rely on honeybee pollination. This percentage accounts for approximately ¼ of the agriculture industry.

230. (C) According to Scientist 2, a ban on neonicotinoids would not be effective in increasing honeybee populations because neonicotinoids are only weakly correlated to honeybee population declines. This weak correlation would lead honeybee populations to continue to decline during a neonicotinoid ban but at a slightly slower rate.

231. (D) Scientist 2 identifies disease and varroa mites as the greatest threats to honeybee health. So it can be inferred that these biotic, or living, factors have the strongest effect on honeybee populations.

232. (A) Most reef-building corals have a mutually beneficial symbiotic relationship with a microscopic unicellular algae called zooxanthellae that lives within the cells of the coral's stomach. The coral provides the algae with a protected environment and the compounds necessary for photosynthesis.

233. (A) As the concentration of CO_2 in the water increases, the pH decreases and the balance between bicarbonate and carbonate shifts increasingly toward bicarbonate as the ocean attempts to buffer the drop in pH by combining H+ with CO_3^{2-} to produce HCO^{3-}. As the carbonate ion concentration decreases, it becomes more difficult for the corals to extract the CO_3^{2-} from the seawater to build their skeletons. It is presently unknown how species vary in their ability to cope with the decrease in carbonate ion concentration in a process known as acclimation.

234. (C) Ocean acidification, like global warming, is a predictable response to those human activities that increase the atmospheric concentration of carbon dioxide. The magnitude and rate of ocean acidification can be predicted with more confidence than the rise in temperature due to global warming, as they are less dependent on climate-system feedbacks.

235. (B) Seasonal changes such as those in temperature and bioproductivity, including variations in photosynthesis and respiration, contribute to fluctuations in ocean pH across all seas. Coastal waters are more likely to be affected by the terrestrial system, such as runoff from rivers, leading to wider variations in ocean pH in these specific geographic regions.

236. (C) Crustaceans live below the saturation point and show a significant increase in calcification, unlike gastropods, corals, and calcareous algae.

237. (A) The aragonite form of calcium carbonate is more soluble than calcite because the aragonite saturation horizon is always nearer the surface of the oceans than the calcite saturation horizon. Therefore, calcifying organisms that produce the calcite form of calcium carbonate (coccolithophores and foraminifera) may be less vulnerable to changes in ocean acidity than those that construct aragonite structures (corals and pteropods).

238. (B) Figure 6.12 shows that bivalve fertility is also negatively impacted by acidification, with slowed development and decreased fertilization and larval survival.

239. (A) As noted in Figure 6.12, the decline in coral calcification is noted as significant with an asterisk.

240. (C) Calcareous algae is most affected, showing a significant decline in growth and photosynthesis, as well as a decrease in reproduction.

241. (D) The outer epidermis provides a layer of protection over the coral animal.

242. (D) Coral eggs combining with sperm in sexual reproduction produces the most genetic diversity.

243. (A) Coral bleaching occurs when the corals lose their color due to stress-induced expulsion of the symbiotic unicellular algae.

244. (A) The best design would be to use two tanks filled with seawater and corals, adding carbon dioxide bubbles to only one tank. The tank with added carbon dioxide is the variable, while the other serves as the experimental control.

245. (B) Any changes in the biological processes in the surface ocean waters also affect the deeper water. This is because organisms and habitats living at the lower levels of the oceans—far from sunlight—rely mainly on the products created by life in the surface waters. On a longer timescale, these organisms may also be vulnerable to acidification and changes in ocean chemistry as higher levels of carbon dioxide mix throughout the oceans.

Chapter 7: Test 7

246. (A) Since the eye's lens is thicker in the middle, it is convex. Since distant objects are being viewed, we can assume that the object is more than two focal lengths away and forms an inverted, smaller image (row 1 in Table 7.1). This image is real because actual light rays are converging to a point after (to the right of) the lens.

247. (D) According to the passage, real images are formed to the right of the lens. According to Table 7.1, all images formed to the right of the lens are inverted.

248. (A) Since the object is located more than two focal lengths from the lens, the first row in Table 7.1 is applicable. A real image forms to the right of the lens and is described as inverted and smaller.

249. (C) According to the passage, convex lenses take parallel rays and converge them to the focal point, which is one focal length from the center of the lens.

250. (A) Since the slide is placed between one and two focal lengths from the lens, the image will be inverted and larger. Because it was placed in the projector upside down, the image on the screen will be upright.

251. (B) Because the image is on the same side as the object, it is a virtual image in one of the two bottom rows of Table 7.1. Since the image is larger, it has to be the convex lens.

252. (B) The only lens in Table 7.1 that produces a same-size image requires the object to be two focal lengths away from the lens (2×30 cm = 60 cm).

253. (C) Since the image of the trees is still distant, the image location is on the same side as the object (to the left of the lens). According to Table 7.1, the only option for smaller images is a concave lens with an upright image.

254. (D) Since the image is located between one and two focal lengths to the right of the lens, the lens must be convex. Table 7.1 indicates that the object is more than two focal lengths from the lens.

255. (C) Since the lens is thicker in the middle, it is convex, and since the candle is within one focal length of the lens, it will form a virtual, upright, larger image.

256. (B) This piece of evidence involves the reduction of predators, which is Scientist A's primary view on the issue. All other choices represent evidence that would be presented by Scientist B regarding the lack of competition for resources from other herbivores.

257. (C) Both scientists agree on the mechanism of the demise of the deer population; their opinions differ on how the population grew and on the data presented.

258. (C) This statement focuses only on the predator-prey relationship, and Scientist B tends to downplay the effects of predator reduction in the description of the Kaibab Plateau.

259. (A) Both scientists agree on the negative implications of interfering in ecosystems.

260. (D) Scientist A focuses on food chain disruption, especially in regard to removing predators.

261. (D) The lesson of this situation, according to Scientist A, is that predators are a necessary part of the ecosystem and removing them can have catastrophic results.

262. (B) The scientists disagree on how the population increased (lack of predation versus lack of competition for resources), but the reason the population decreased catastrophically is clear—disease and starvation from overpopulation.

263. (D) Scientist B believes that the reduction in competition for grasses caused the increase in the population.

264. (B) While the other statements could potentially be true, this cautionary statement shows a link between the practice suggested for increasing the duck population and what happened on the Kaibab. Removing the midsized mammals (raccoons, foxes, and so forth) in this ecosystem could have a similar ripple effect that removing the large predators (such as coyotes) had on the Kaibab Plateau.

265. (B) Although the two scientists differ in their reactions to the environmental mechanisms that led to the problem, they agree that a lesson should be learned from the Kaibab Plateau situation about human intervention in natural ecosystems.

266. (B) To determine the effect of amplitude, tension must remain constant so it is a controlled variable.

267. (D) In Figure 7.3, the high-tension/high-amplitude wave traveled approximately 1.10 m at the one-second mark. Therefore, it traveled 1.10 m/s.

268. (C) The straight line indicates that it traveled an equal distance each tenth of a second as time went on, thus remaining at a constant speed.

269. (A) When comparing the two high-amplitude lines, the steeper slope for the high-tension case shows that the wave traveled a greater distance each second than it did in the low-tension case.

270. (C) Since the slopes of the two low-tension cases were almost identical but the amplitudes were different, one can conclude that amplitude has no significant effect on wave speed.

271. (A) One can infer that loud sounds are big waves of large amplitude, but amplitude has no effect on wave speed.

272. (A) The only time wave speed changed in Figure 7.3 was when the tension was increased, which effectively altered the characteristics of the medium. Thus one can infer that the speed of light will be altered by the material through which it travels.

273. (D) Since the waves maintained a steady speed as they traveled, one can infer that a water wave will maintain constant speed.

274. (C) The low-tension/low-amplitude wave went about 0.8 m in one second. Since its speed was constant, it should move six times that distance in six seconds, thus moving approximately 4.8 m.

275. (C) Since wave speed is independent of the size of the wave (amplitude), it should take the same amount of time for the small wave to travel the same distance as the large wave.

276. (B) Heptane is an alkane (circle marker) with a molar mass of 100 g/mol. In Figure 7.4, the corresponding dot has a boiling point just less than 100°C.

277. (B) As the molar mass for both alkanes and alcohols increases, so does the boiling point. This is shown by the two linear trends with positive slopes in Figure 7.4.

278. (C) Propanol is an alcohol, so it follows the trend of the squares at the top of Figure 7.4. Extrapolating backward from a molar mass of about 74 g/mol to 60 g/mol would correspond to a 12°C to 14°C decrease in boiling points.

279. (C) Since 150°C is higher than the boiling point of pentanol, the substance would already have boiled and become gaseous.

280. (A) The addition of a hydroxyl group to any of the alcohols makes the boiling points higher than those of the alkanes that have similar molar masses. Look at pentane and butanol: these have similar molar masses, but butanol (which has a hydroxyl group) has a much higher boiling point.

281. (C) This model has the six carbon groups and the hydroxyl group that makes it an alcohol. In Table 7.2, all substances ending in *-ol* are alcohols. The chemical hexane also has six carbon groups, so the base of the structure must have something to do with the *hexa-* prefix.

282. (B) The model shows that each point on the skeletal structure indicates a carbon atom. The carbon on each end is filled with three hydrogen atoms and the carbons in the middle each have two hydrogen atoms attached. Eight lines mean 10 carbons: 8 inside ($8 \times 2 = 16$) and 3 on each end ($2 \times 3 = 6$) equals 22 hydrogen atoms ($16 + 6 = 22$).

283. (B) To figure this out, extend the lower line on Figure 7.4 out to 142 g/mol. The boiling points of alkanes seem to increase about 60°C for an increase of 25 g/mol. A molar mass increase from 100 g/mol to 142 g/mol equates to about a 90°C increase in the boiling point, which is closest to 174°C.

Chapter 8: Test 8

284. (A) Figure 8.1 best represents the relationship between the dependent and independent variables. Figure 8.2 is a bar graph (which is best used for categorical data) and does not have the correct data. Figure 8.3 has the dependent and independent variables on the wrong axes and also does not represent the x-axis data in an appropriately scaled manner. Figure 8.4 does not correlate the correct data, as a downward curve should immediately be expected when looking at the results from the experiment.

285. (A) This is the only question that could be addressed using only the data students have already collected.

286. (C) No relationship was shown between the cost of sunglasses and the amount of UVB they blocked.

287. (C) If each bottle of sunscreen held 10 oz, the costs would be \$0.89, \$0.49, \$0.79, \$1.59, and \$1.09 per ounce. SPF 4 can be eliminated because it is more expensive than SPF 8 and clearly blocks less UVB. SPFs 30 and 50 can also be eliminated because although they block more UVB than SPF 8, the higher cost (double and triple the cost per ounce) could not possibly offset the only marginal increase in protection.

288. (D) It would be important for students to control the amount and thickness of the sunscreen to obtain consistent and reliable results. The other selections represent variables that cannot be controlled, such as sunscreen ingredients, or are not pertinent to this investigation.

289. (D) This cannot be predicted because there is no clear relationship between the cost of the glasses and the amount of UVB blocked.

290. (A) If the trend from Table 8.2 continues, the amount of UVB reaching the sensor will decrease slightly.

291. (B) Students are testing the type of sunglasses. The experimental, or independent, variable is the item that is changed between each trial.

292. (A) As the SPF increases past 30, only slight increases in the amount of UVB blocked occur.

293. (C) This is the only question that can be answered using the existing equipment. Students can use a consistent SPF and vary the thickness to determine whether this affects what reaches the sensor.

294. (D) This statement is false. The more water in the initial amounts, the *lower* the evaporation rates, although this is probably a result of the different surface areas exposed and not a function of the amount of water. The data for water alone do not warrant any conclusion here.

295. (B) 80.0 mL – 3.0 mL = 77.0 mL.

296. (B) Since rubbing alcohol had the highest evaporation rate, Student 1's hypothesis is supported. The average evaporation rates for Student 2's data are approximately the same for the three liquids. Student 3's hypothesis is disproved because the data show that the greater the surface area, the greater the evaporation rate.

297. (D) The data in Table 8.3 indicate that the amount evaporated is directly proportional to surface area. The average amount evaporated goes up approximately 9 mL for every 4 cm^2 of surface area exposed. Since the surface area is increased by a factor of 5, the amount evaporated will also (5×9 mL = 45 mL).

298. (A) If you extrapolate the rubbing alcohol line for two additional weeks, it will reach zero at approximately 9 weeks of exposure.

299. (A) The line for water drops 21 mL in seven weeks. That's an average evaporation rate of 3 mL of water each week.

300. (C) The slope of the line for rubbing alcohol in Figure 8.5 is constant over the seven weeks. This indicates that the rate of evaporation is constant.

301. (C) The data indicate that none of the vegetable oil evaporated. Therefore the graph would be a steady, flat line at 80 mL.

302. (A) The variation between the trials for rubbing alcohol is greater for the experiments with bigger surface areas.

303. (D) The initial amount of liquid was the same in each case for rubbing alcohol.

304. (D) To answer this question, one can examine the last five rows of Table 8.4, where the radius and release height are constant, but the mass is changing. Notice that the speed remains constant.

305. (C) When mass and radius are held fixed in the middle rows of Table 8.4, the kinetic energy is one-tenth of the release height.

306. (A) The middle rows on Table 8.4 have a fixed mass and radius but differing release heights. One can compare the effect of doubling the release height from 0.05 m to 0.10 m, or 0.10 m to 0.20 m, on gravitational energy to observe the doubling effect.

307. (D) All the gravitational energy at the top of the swing transfers to kinetic energy at the bottom because the values are identical in Table 8.4.

308. (A) According to the data in Table 8.4, when the mass doubles from 0.020 kg to 0.040 kg, the kinetic energy also doubles.

309. (C) The middle rows in Table 8.4 are for a fixed pendulum design, and the speed increases by a factor of 1.4 J when the kinetic energy doubles.

310. (A) The last five rows in Table 8.5 (where radius and release height are held constant) demonstrate that centripetal acceleration does not depend on mass.

311. (C) The first five rows of Table 8.5 (where mass and release height are held constant) show that each time the radius doubles, the centripetal force values are cut in half.

312. (D) By examining the middle rows in Table 8.4, where mass remains the same, when the speed doubles from 0.99 m/s to 1.98 m/s, the kinetic energy quadruples from 0.005 J to 0.020 J. Thus, kinetic energy is directly proportional to the square of the speed.

313. (B) The last five rows of Table 8.5 apply to a pendulum with a 0.40-m arc and a release height of 0.25 m. The centripetal force increases by 0.123 N for each 0.010 kg of mass. Adding 0.123 N to the last row of Table 8.5 yields 0.613 N + 0.123 N = 0.736 N.

314. (B) The target for the North Lake area is 25 μg/L. In all years of the study, the values of phosphorus measured in North Lake have exceeded that level.

315. (B) According to Table 8.6, a phosphorus level between 12 μg/L and 24 μg/L would be classified as mesotrophic. The target for Bay Area 2 is 14 μg/L.

316. (D) South Lake is the only area that shows a downward trend in phosphorus levels starting in 2005.

317. (C) In 2002, the South Lake phosphorus level was 38 μg/L. This falls in the category of eutrophic.

318. (D) Bay Area 2 fell in the range of 12 μg/L to 24 μg/L for 19 years of the study.

319. (D) Bay Area 1 fell in the range of 12 μg/L to 24 μg/L during 2011.

320. (A) By subtracting the approximate value of the shortest bar from that of the highest bar in Figure 8.10, a range can be determined (61 – 21 = 32 μg/L).

321. (A) Bay Area 1 had a range of only 5 μg/L (13 – 8 = 5 μg/L). It is important to pay attention to the scale in each figure, as each has a different unit scale on the y-axis. The ranges for each lake area can be found by subtracting the approximate value of the shortest bar from that of the highest bar in each figure.

322. (B) According to the passage, the value must be at or *below* the target value of phosphorus to surpass the standard. By counting the bars at or below 10 μg/L in Figure 8.7, one can determine that the standard was met or surpassed for 10 years in Bay Area 1.

323. (B) In 1993, Bay Area 1 had an approximate phosphorus concentration of 9 μg/L and North Lake had a concentration of 50 μg/L (50 – 9 = 41 μg/L).

324. (A) In 2005, Bay Area 1 had the most similar phosphorus concentration to that of the neighboring lake.

325. (D) To be classified as hypereutrophic, the phosphorus level of a lake needs to be higher than 96 μg/L. None of the lake areas exceeded that level during the 21-year study.

326. (C) The North and South Lake areas most frequently fell into the range of 24 μg/L to 96 μg/L.

327. (B) According to the passage, lakes classified as eutrophic have low levels of dissolved oxygen. In 2010, North Lake was the most eutrophic of all the areas studied and would therefore be expected to have the lowest dissolved oxygen level.

Chapter 9: Test 9

328. (A) The independent variable is the variable the experimenter purposely changes. The types of metals were varied throughout this experiment, so this is the independent variable.

329. (B) The control variable must remain constant for all trials. The type of metal cannot be controlled since it is varied for each experiment. Independent variables cannot be control variables.

330. (B) Silver acts as the cathode in this reaction. From the passage, we know that metal ions (Ag^+) join the Ag strip as solid metal atoms. This diminishes the amount of metal ions in the jar. After repeated use, the silver ions will be attracted to the silver cathode, and the concentration of Ag^+ will decrease.

331. (C) To get a positive voltage, the sum of the cathode and anode must be 0.63. This is only true if zinc is the anode with 0.76 V and lead is the cathode at –0.13 V.

332. (D) The passage stated that metal from the anode strip dissolves into solution and metal from the solution is added to the cathode strip. Zinc metal was the anode in Experiment 2, so its atoms dissolved into zinc ions, decreasing the strip's mass. Copper was the cathode, so the dissolved zinc ions attached themselves to the strip, increasing the strip's mass.

333. (A) Figure 9.1 shows zinc as the anode and copper as the cathode. The anode table in Table 9.2 shows that zinc has a more positive voltage than copper. It can thus be inferred that aluminum would be the anode, because it also has a more positive voltage than copper.

334. (D) The aluminum anode has a potential of 1.66 V, and the copper cathode has a potential of 0.34 V. Added together, this yields a sum of 2.00 V. Switching the anode and cathode would give a sum of –2.00 V, which is not a choice.

335. (C) Each copper/zinc cell can produce 1.10 V. Using the equation $12 \div 1.10 = 10.9$ V, or between 10 and 11. Ten cells clearly do not produce quite enough voltage, so 11 cells are needed.

336. (A) In Table 9.3, the voltage divided by the resistance predicts the current.

337. (D) The passage states that power is the product of current and voltage, so $150 \text{ V} \times 3.0 \text{ A} = 450 \text{ W}$.

338. (C) For the 120 V rows in Table 9.3, when the resistance triples from 120 ohm to 360 ohms, the power drops from 120 W to 40 W, thus decreasing to one-third of its initial power.

339. (A) For the 480-ohm rows in Table 9.3, when the voltage doubles from 120 V to 240 V, the current doubles from 0.25 A to 0.50 A.

340. (D) According to Table 9.3, the 480-ohm bulb at 120 V has an energy usage of 108 kJ per hour. To find the energy needed, we use $108 \times 4 = 432$ kJ.

341. (B) According to the data in Table 9.3, doubling the resistance from 120 ohm to 240 ohm decreases the power to one-half of 120 W. Tripling the resistance decreases the power to one-third of 120 W. Thus, power is inversely proportional to resistance (for a fixed voltage). When resistance is increased by a factor of 5 from 120 ohm to 600 ohm, the power must reduce to one-fifth of 120 W, or 24 W.

342. (A) In Table 9.3, a basic circuit with a single 240-ohm bulb has a power output of 60 W when connected to a power supply of 120 V. According to Table 9.4, when identical bulbs are connected in parallel at 120 V, they also output 60 W. According to the passage, power is the amount of energy transferred each second.

343. (C) According to Table 9.4, as the number of bulbs in a series circuit increases, the individual bulbs output less power.

344. (A) According to Table 9.3, increasing the resistance of a single bulb in the circuit decreases the power output. Since the circuit's power output increases as bulbs are added in parallel in Table 9.4, one can infer that the resistance of the entire circuit decreases.

345. (B) This statement is false because even if Bulb 2 breaks, there is still a series pathway for Bulbs 1 and 3 through the power supply.

346. (D) According to Table 9.4, more bulbs being added to a parallel circuit results in more power output from the entire circuit.

347. (C) The five bulbs that went out are wired in series with each other because when one went out, the others did as well. Since the other four bulbs remained lit, they must be on a separate circuit.

348. (A) According to Table 9.4, the current decreases as bulbs are added in series. Table 9.3 shows that current decreases in the circuit when resistance increases. Thus, adding bulbs effectively increases the total circuit resistance.

349. (D) According to Table 9.4, three 240-ohm bulbs wired in series to a 120 V power supply outputs a total of 20 W. That's 20 J of energy transferred each second, so in 10 seconds, there is a total of 200 J.

350. (C) According to Table 9.4, each bulb wired in parallel requires 60 W, so for six bulbs, the equation is 6×60 W $= 360$ W.

351. (B) Experiment 1 looked at cathode rays that were attracted to a positively charged area. This suggests that the rays are negatively charged. Experiment 2 suggests that something inside the atom is sturdy or dense enough to reflect alpha particles.

352. (D) If the mass-to-charge ratio of a particle is small, either the mass of the particle is small or the charge of the particle is large. From the ratio alone, it is impossible to know which factor creates the small ratio. Both (A) and (C) might be true, but (D) is the best conclusion because it notes both possible factors.

353. (C) The neutron has some mass but is neutral, so it has no charge. An object with zero charge would have an undefined mass-to-charge ratio, no matter what the mass of the particle is.

354. (C) Both experiments agree that atoms have subatomic particles. This conflicts with Democritus's view that atoms are indivisible.

355. (D) Only 1 out of every 20,000 particles was deflected to a large degree, meaning that 19,999 out of every 20,000 particles went straight through the foil. This suggests that there is a tiny chunk of dense material in an atom, but the rest is empty space.

356. (D) The gold foil was 8.6×10^{-8} m, or 0.000000086 m. One micrometer is 0.000001 m, but the gold foil is not as thick as 1 μm. The only reasonable answer is 0.086 μm because it shows that the gold foil is much thinner than 1 μm.

357. (A) This statement is talking about cathode rays that have been proven to be negatively charged and very small.

358. (C) This headline matches what the gold foil experiment concluded. An atom must have a dense particle in it to make the alpha particles bounce off the foil. However, this particle must be very small if only 1 out of 20,000 alpha particles bounced back.

359. (B) In the capture theory, the moon was formed in another area of the solar system; therefore, it would likely have different rock compositions.

360. (C) In the fission theory, the moon came directly from the earth and would therefore have the same rock composition. In the coaccretion theory, the moon and the earth formed side by side from the nebula and would therefore have very similar compositions.

361. (B) The fact that Jupiter and Saturn both have captured moons makes it more likely that the earth could have captured a moon.

362. (A) Fission = Daughter (the moon was "born" from the earth in this theory). Coaccretion = Sister (the moon and the earth formed side by side from the same "parent" material). Capture = Spouse (the moon was formed elsewhere and was "married" to the earth when captured by the earth's gravitational field).

363. (A) This is a description of fission theory, the leading theory in the mid-1930s. This is most evident by the mention of the sun's gravity and the phrase "a bulge broke away with such momentum that it could not return to the body of Mother Earth."

364. (D) If the moon was formed 4.5 billion years ago, it could not have broken away from the earth to form the Pacific Ocean basin.

365. (C) In the capture theory, the moon formed in another part of the solar system; therefore, the earth's gravity would not have played a role in the formation of its crust.

366. (D) The giant impact theory is the only theory to account for why the moon rocks partially, but not fully, match the rocks on the earth. In this theory, the collision with the Mars-sized object would have caused a mixing of materials from both.

367. (D) Based on the information in the passage, it is evident that the giant impact theory is the most current theory of moon formation based on scientific evidence.

368. (C) These two pieces of evidence best support the coaccretion theory, in which the earth and moon formed side by side.

369. (A) All of the other pieces of evidence point toward a moon formed close to the earth, but this theory does not account for the similarity in the crust/mantle rock from both the moon and the earth.

Chapter 10: Test 10

370. (B) The velocity in the first column of Table 10.1 always increases by 9.8 m/s each and every second.

371. (C) According to Table 10.1, the distance fallen in the first second is 4.9 m. During the second second, the distance fallen is 19.6 – 4.9 = 14.7 m. The intervals only increase from there.

372. (B) During the fourth second, the air drag force increases from 0.061 N to 0.080 N, an increase of 0.019 N.

373. (D) Since the velocity goes up about 10 m/s each second, we can add 10 to 118 m/s (the speed 12 seconds into the fall) two more times to get 138 m/s at the 14-second clock reading.

374. (A) The velocity shown for a ball falling with air increases at a lesser rate than the velocity of one falling without air (free fall).

375. (D) The last column in Table 10.1 shows big changes each second initially, but by the end of the drop, the air drag does not change.

376. (D) Figure 10.1 shows that the 30 g mass gains speed at a greater rate than the 10 g mass (both falling with air).

377. (B) During the first two seconds in Figure 10.1, there is not much difference between the with-air and without-air data. Beyond two seconds, when the velocities are greater than 20 m/s, the data diverge significantly.

378. (D) Figure 10.2 shows the 0.05 cm ball's velocity plateauing at about 62 m/s at 12 seconds into the fall.

379. (A) In Figure 10.2, the line for both balls falling without air is identical.

380. (C) In Figure 10.1, the 10 g ball reaches terminal velocity 6 seconds into the drop, whereas the 30 g ball reaches terminal velocity 12 seconds into the drop.

381. (B) The belly-first twin hits more air just like the larger radius ball. By examining Figure 10.2, one can see that the larger radius ball reaches terminal velocity sooner than the smaller radius ball, and the belly-first twin will do the same.

382. (C) According to Figure 10.1, both the 10 g ball and the 30 g ball have the same velocities when falling without air. Therefore, mass has no effect on free-fall velocities and the balls will reach the ground at the same time.

383. (A) According to Figure 10.1, the greater mass falls with greater velocity through the air and will get to the ground sooner.

384. (B) Both figures and Table 10.1 show that dropped balls speed up steadily. There is no evidence that the steady gain in velocity will ever change.

385. (B) The mass of sample can be found by taking the mass of the crucible + sample and subtracting the mass of the empty crucible ($32.569 - 26.449 = 6.120$ g).

386. (C) Even though the sample was heated again, no additional water molecules were driven off between the third and fourth heatings; instead, only anhydrous $MgSO_4$ was left in the container.

387. (A) *Percent mass* is the mass of the investigated quantity (mass of water) divided by the total mass. The ratio of these numbers describes the percent of water in the original solution.

388. (C) The difference in the first two rows of Table 10.2 shows that approximately 6 g of the hydrated sample were used. The difference between the mass of the sample at the end and the mass of the sample before heating was approximately 3 g (of water driven off). This indicates the sample was around 50% water and 50% magnesium sulfate.

389. (A) The percent mass of water would increase if some solid splattered out of the jar. When the mass of the sample was measured after the last heating, it would be less than expected because some of the solid had sprayed out, but the student would (incorrectly) assume the lost mass to be solely water that had evaporated.

390. (C) The mass of the sample increased during the additional 40 minutes. The number of $MgSO_4$ crystals could not increase, and any "expansion" in size would not change the mass. A change in temperature also does not change the mass. The only reasonable choice is that water molecules from the atmosphere were able to rehydrate the crystals.

391. (B) The hydrate would still heat up and water would be driven off, but that water could not leave the crucible. If the crucible were opened, drops of water would be found on the sides, but the mass would not have changed because the water was not allowed to go into the atmosphere.

392. (B) Both beakers contain 50 g of measured solid, but much of that mass would be water molecules in Beaker 2. The 50 g of solid in Beaker 1 contains all $MgSO_4$ and no water molecules. When dissolved, there would be a much higher concentration of Mg in Beaker 1.

393. (D) Hyperion's 21.3-day orbit is closest to the moon's 27.3-day orbit.

394. (B) Andrastea and Miranda have the same orbital radius, and they orbit planets with different mass. Thus, the effect of only one variable (central planet mass) may be analyzed.

395. (D) The data for Saturn show that larger radii correlate with greater periods. The moons of Jupiter and Uranus confirm that observation.

396. (A) The data indicate that gravitational acceleration is directly proportional to the inverse radius squared. In other words, gravitational acceleration is inversely proportional to the square of the radius.

397. (D) Larger radii correlate with smaller orbital speeds. Since 422 million meters is less than 181 million meters for Amalthea, than the correct answer must be significantly less than Amalthea's orbital speed of 26.5 km/s. Thus, 17 km/s is the only answer that fits that criterion.

398. (A) Andrastea (orbiting Jupiter) has about 20 times the gravitational acceleration of Miranda (orbiting Uranus), but they have the same orbital radius. Jupiter has about 20 times the mass of Uranus, so gravitational acceleration appears to be directly proportional to planet mass. Using Figure 10.3, the greater slope for Jupiter compared with that of Saturn confirms this observation, because Jupiter is about three times the mass of Saturn.

399. (C) The passage indicates that speed was calculated by taking orbital circumference divided by period and that gravitational acceleration was calculated by the square of speed divided by orbital radius.

400. (C) We divide Phoebe's period by that of the moon ($550 \div 27.3 = 20$).

401. (C) According to Table 10.4, Helene travels 10.0 km/s. That means in 15 seconds, it will travel 150 km.

402. (D) As noted earlier, since gravitational acceleration is directly proportional to planet mass, a light planet like Mars would have moons whose gravitational acceleration was much less than those of Jupiter or Saturn, so the line would be below that of the moons of Saturn.

403. (A) The temperature changed 10°C during the first 6-minute interval. From 6 to 20 minutes, the temperature barely decreased. The temperature did decrease from 20 to 30 minutes, but only about half as rapidly.

404. (A) If the thermometer was touching the bottom of the test tube, it may have recorded the temperature of the glass as well as the temperature of the sample. This would have caused the temperature to be higher. A reasonable thermometer reading would be 80°C if the tip of the thermometer was resting on hot glass and the rest of the thermometer was in the 55°C solution.

405. (D) There are three separate parts to this cooling curve. The first and third parts are downward-slanting lines that indicate the temperature decreased in both parts. The middle part shows constant temperature, so the phase must have been changing.

406. (B) The middle plateau shows the lauric acid cooled down but did not change temperature. This must indicate a phase change. Liquid to solid is the freezing point.

407. (A) Since 25°C is well below the freezing/melting point of lauric acid, all material would be in the solid phase.

408. (B) As it cooled, the temperature dropped 9°C in the last 10 minutes of the experiment. To cool from 34°C to 25°C (another 9°C), it would take another 10 minutes.

409. (A) The melting/freezing point does not depend on the amount of lauric acid. It may take more time to reach the melting point, but it would still occur around 43°C.

410. (C) Since 50°C is higher than the melting point of lauric acid, the sample would never have crystallized into a solid. It would have remained a liquid throughout the experiment.

411. (B) Since 25°C is between the melting and boiling point, all of the substance would be a liquid. This is equivalent to water at 25°C, which is between the melting point of ice and the boiling point of water.

412. (A) If the water started at 55°C and was allowed to cool to 20°C, it would indeed decrease in temperature, but it would not go through melting or boiling. There would be no plateau areas in Figure 10.5, but the temperature would decrease.

Chapter 11: Test 11

413. (C) Volume is a measurement of how much space something occupies. Sample 6 had a volume of 44.8 L, which was the largest volume of any sample.

414. (D) Samples 1 and 3 have the same volume, which means they are the same size. Sample 3, however, has a mass of 10 g, whereas Sample 1 has a mass of only about 1 g.

415. (C) Both samples are 22.4 L. Avogadro's law states that any two gases at the same volume, temperature, and pressure will contain the same number of molecules.

416. (C) An 11.2 L sample of oxygen gas would have a mass of 16.0 g. To calculate the mass for 22.4 L, one needs to double the volume, which would be twice the mass of Sample 7.

417. **(B)** The neon has a mass of 10.1 g, and 11.2 L of hydrogen gas has a mass of 1 g. One would need 10 times that much mass of hydrogen gas, or $10 \times 11.2 = 112.0$ L.

418. **(D)** Avogadro's law states that equal volumes of gases at the same pressure and temperature would have an equal number of molecules. All three samples have a volume of 11.2 L.

419. **(B)** While 11.2 L of oxygen has a mass of 16 g, 22.4 L of helium has a mass of only 4 g. Therefore, 11.2 L of helium would have a mass of 2 g. Thus, a sample of oxygen gas is 8 times heavier than an equal-sized sample of helium gas.

420. **(A)** If 11.2 L of oxygen has ½ mol of molecules and a mass of 16.0 g, then 1 mol of molecules would be twice the volume and twice the mass to equal 32.0 g.

421. **(C)** Chlorophyll *a* and *b* have the lowest absorption in the 525 to 625 nm portion of the spectrum, which corresponds to the greatest reflection. Even without a knowledge of light principles, this can be determined through the process of elimination, since the other answers do not match the data in Figure 11.1.

422. **(B)** The blue portion of the visible spectrum is between 450 and 495 nm.

423. **(B)** A wavelength of 440 nm represents considerable absorption for all three pigments.

424. **(A)** From Figure 11.1, it can be determined that red light has wavelengths greater than 620 nm. Chlorophyll *a* has the highest peak in the range of wavelengths greater than 620 nm.

425. **(D)** Carotenoids absorb 50% of visible light at 450 nm, and chlorophyll *b* absorbs 80%, which means the carotenoids have ⅝ (62.5%, or approximately 60%) of the absorbing power at that wavelength.

426. **(A)** The pigment with the highest absorption at 425 nm is chlorophyll *a*.

427. **(B)** The highest peak for chlorophyll *b* in Figure 11.1 is at 450 nm.

428. **(C)** Absorption and reflection are opposites, but one would not have to be familiar with the term or process of reflection/principles of color to determine that green light is reflected by chlorophyll *a* and *b*, making plants appear green. Figure 11.1 shows that (A) and (B) are false. (D) is the opposite of (C) and can be eliminated by looking at Figure 11.1, which shows higher absorption in the red portion of the spectrum than in the green.

429. **(C)** The peaks of this pigment are the opposite of those of chlorophyll *a* and *b*, as it absorbs the most light in the green portion of the visible spectrum where the absorption of chlorophyll is the lowest.

430. **(C)** Particles tend to move in straight lines. As the particles stream from the bulb in straight lines, they are blocked at angles by the card and form the large shadow on the screen.

431. (D) The first three answers all deal with the reflection, refraction, and absorption of light waves and support Huygens's theory. Visualizing a photon as a bundle of energy is a particle-like concept.

432. (A) The vacuum of space has no air, so how can a wave travel without matter to transfer the energy? Huygens's theory could not explain this, whereas Newton's particles could travel through the nothingness of space.

433. (A) The bright regions on the screen can be explained by constructive interference and the dark areas by destructive interference. Interference is a wave phenomenon.

434. (D) The fuzzy edge may be explained by the diffraction (spreading) of light around the sharp edge of the penny. The bright area at the middle of the shadow is the result of constructive interference of the light diffracting around the edge of the penny.

435. (C) The semicircles are like water-wave ripples traveling through the air. This supports Huygens's wave theory.

436. (B) Since the quote states that light does not consist of any "transport of matter" (arrows or particles), it is not consistent with Newton's particle theory of light. The emphasis of the spreading of light does not contradict Huygens's wave theory.

437. (A) The photon as a basic unit of light can be visualized as a particle, and this is consistent with Newton's particle theory.

438. (D) If light is a particle moving in a straight line, rotating the second filter should affect the intensity of the light.

439. (A) This high absorbance in Figure 11.4 means lots of light is being absorbed. $NiSO_4$ has an absorbance of around 0.075 at 380 nm. This light is purple in color. Most of the purple light gets absorbed by the green nickel sulfate solution.

440. (B) Low absorbance means little light is absorbed and lots of light passes through. The green $NiSO_{4(aq)}$ solution allows most of the green light from the spectrometer to pass through, but it absorbs most other colors.

441. (D) As the concentration increases, the absorbance increases proportionally. The concentration of $NiSO_{4(aq)}$ is changed, and this changes the absorbance proportionally.

442. (C) The absorbance of $NiSO_{4(aq)}$ increases about 0.90 for each increase of 0.08 mol/L. This is a proportional increase that can be used to predict the concentration of unknown solutions.

443. (C) Figure 11.4 shows that $NiSO_4$ does not absorb much light at 490 nm. Since the absorbance is so low, the absorbance of each sample in Investigation 2 will be considerably lower than the absorbance found at 740 nm.

444. (B) A solution of NaCl is a clear liquid, so it will not absorb any light. All colors of light will pass straight through it. Spectrometry depends on certain frequencies (colors) of light being absorbed.

445. (D) Following the trend, a 0.48 mol/L solution should have an absorbance of 0.542. However, the fingerprint would cause more light to be absorbed, so the answer has to be greater than 0.542.

446. (A) The chemical $NiSO_{4(aq)}$ has a unique spectral fingerprint that is shown in Figure 11.4. A more concentrated solution would absorb more light because more of the chemical would be present. The graph would be the same as the one in Figure 11.4, but the line would be higher, showing that each wavelength would have slightly more absorbance.

447. (B) The data for Group 1 show *S. aureus* growing most rapidly and to the highest turbidity at 37°C.

448. (C) At 600 minutes, the data reached their maximum value. Visualizing a graph of this data, one would see it level off at this point, corresponding to the stationary phase of growth.

449. (D) pH is one of the variables that is controlled for Group 1 and is the variable being tested by Group 2.

450. (B) Twenty-nine minutes most closely approximates the generation time. One can choose any two pieces of data up to 100 minutes where the absorbance value doubles (for example, 0.047 and 0.099) and determine the amount of time that has passed by subtracting the time readings (68 and 97 minutes, respectively). Even if other data points are selected (such as 0.035 and 0.070) or one chooses another method of approximation, (B) is the only answer that is reasonable.

451. (B) The temperature of incubation is the experimental variable for Group 1, not one of the controls.

452. (A) *S. aureus* has a faster growth rate at pH 7 than at pH 5, as evidenced by the more rapid increase in absorbance values in Table 11.5.

453. (B) The information in the passage provides the distinction between acidity and alkalinity. At this high level of alkalinity, *S. aureus* bacteria cannot grow, as evidenced by the lack of increase in turbidity.

454. (C) Group 1/Trial 3 and Group 2/Trial 3 were both performed under conditions of 37°C and pH 6.

455. (A) It can be concluded from the growth data that 37°C and pH 6 represent the ideal growth conditions for *S. aureus*.

456. (A) The graph will be a curve, as the data increase and then level off between 10 and 24 hours.

457. (C) This trial is the only one mentioned that occurred at 37°C and featured growth that was retarded but not completely inhibited by the conditions.

458. (A) Time is the independent variable, as its presence determines the value of the other variables.

459. (C) Comparing generation time at a standard temperature would be the most accurate way to determine the type of bacteria using spectrophotometry. Although (A) and (D) are true of *S. aureus*, they are likely true of many other bacteria as well and lack the specificity of a calculation such as generation time.

460. (B) The population (as indicated by turbidity) remained relatively stable, which means that birth and death must be balanced.

Chapter 12: Timed Test

461. (D) The ionization trends upward briefly but then drops severely between beryllium (Be) and boron (B). After that, the trend is generally toward higher ionization levels.

462. (D) The formula refers to the fourth ionization energy of nitrogen as nitrogen is going from the +3 charge to the +4 charge.

463. (A) Oxygen is shown to have eight electrons. Oxygen is neutral when it has eight electrons (because it has eight protons), so the image shows the first ionization.

464. (B) The last two ionization energies are always significantly larger than any of the others. This means the last two electrons are very hard to remove because they are closest to the nucleus in the inside shell.

465. (C) Each piece of helium has two electrons. To remove all of the electrons from helium, both electrons must be removed from 1 mol of particles. This requires the amount of energy equal to the sum of the first and second ionization energies ($2,372 + 5,250 = 7,622$ kJ).

466. (C) Brand P sustained a voltage of 2.2 V or more for approximately 3.5 hours.

467. (B) The data for the alkaline batteries are nearly identical, whereas the data for the heavy-duty batteries are quite different.

468. (D) The uncertainty in the data set for the remote-control car is so significant that no valid conclusion may be made about battery performance. Most likely, the method of testing the car was flawed.

469. (C) Below the 2.2-V line, the slope for Brand E is steeper than that for Brand P.

470. (B) If the batteries had been used previously, then any experiment to test longevity would be invalid. Effects of temperature and manufacture date may have only a slight effect on the data. The selection of time intervals is arbitrary, although they must be small enough to discern performance.

471. (B) According to Table 12.2, the LED times are about four times as long as the incandescent bulb times.

472. (A) Scientist 1 mentions that research is currently being conducted to detect the dark matter particle. Scientist 2 complains that many years of research have so far been fruitless. Both scientists agree that the particle has yet to be detected.

473. (A) Since Neptune was unseen matter that affected the orbit of Uranus, it was essentially dark matter, a concept that Scientist 1 supports wholeheartedly.

474. (A) The excessive gravitational lensing around galaxies indicates that the galaxies are more massive than expected. This supports Scientist 1's theory of dark matter.

475. (B) Scientist 2's last statement in the passage emphasizes his or her belief that Einstein's theory of gravity needs to be modified.

476. (C) The measured values in the graph are from experimental observations and are shown with the solid line. This line shows that the orbital speeds of the stars remain fairly steady as orbital radius increases.

477. (B) Vera Rubin's comment indicates that she is more comfortable with modifying the current laws of gravity rather than believing there are unseen particles of matter in the universe. Scientist 2 also believes we need to modify the current laws of gravity.

478. (D) Since it took almost 50 years to discover the Higgs boson, one would expect that it may take at least that long to detect the dark matter particle. Scientist 1 is hopeful that we will detect that particle in the future.

479. (D) According to Table 12.3, it takes the 2.00 kg mass 1.85 seconds to complete one period. The passage defines a period as the time to complete one complete vibration. Thus, 10 complete cycles will take 10 times as much time as one period, or 18.5 seconds.

480. (D) Spring stiffness is changed on purpose in Group B's experiment, which makes it an independent variable. The other three variables listed must remain unchanged (controlled) for it to be a valid experiment.

481. (C) According to Table 12.3, the period increases with mass. Notice that the period increases the greatest amount for the first increase in mass, and the increase gets less as the mass increases equally each time.

482. (D) Group C's periods were all about 4 seconds. The spring stiffness that corresponds with that period in Group B's data is Trial 5 with a spring stiffness of 25.0 N/m.

483. (B) Group B's data decrease at a diminishing rate with spring stiffness. Extrapolating that trend in the data table for two more rows indicates that 3.2 seconds is the only reasonable choice.

484. (A) Group A's data show that increasing the mass significantly increases the period. Group B's data show that increasing spring stiffness decreases the period significantly. Group C's data show that stretch distance has no significant effect on the period. Option A is the only choice consistent with these conclusions.

485. (C) Figure 12.5 shows that both the inner and outer radius of the habitable zone increase with start temperature. The solid line for the outer radius, however, increases at a greater rate, causing the width of the habitable zone to increase with star temperature.

486. (B) The passage clearly states that the distance at which a planet orbits its host star is the most important factor that affects a planet's surface temperature.

487. (C) Using Figure 12.5, one first draws a vertical line at the star temperature of 6250 K. That line intersects the dashed inner radius line at 1 AU and then intersects the solid outer radius line at about 2.2 AUs. The habitable zone, supporting life, is between the inner and outer radius.

488. (C) Using Figure 12.6, one can see that the dashed line for Kepler 426-B blocks its sun periodically every 0.6 years. Drawing a horizontal line on Figure 12.7 at 0.6 years, one can see that it intersects the curve at a distance from the sun of 0.7 AUs.

489. (B) Using Figure 12.7, one sees that Planet 1's period of 0.6 years corresponds to a distance from its sun of 0.7 AUs. Examining Figure 12.6 at a start temperature of 5500 K, 0.7 AUs falls in the habitable zone between the inner and outer radius. Following the same approach, one sees that Planet 2's period of 1.8 years corresponds to a distance from its sun of 1.5 AUs, which falls outside the habitable zone, which will not support liquid water.

490. (B) Examining Group A's frequencies in Table 12.6, a 349 Hz frequency lies between Pipe 3 and Pipe 4's frequencies. Thus, the length of the pipe must be between 0.52 m and 0.44 m. The only choice in this range is 0.46 m.

491. (D) Choices A, B, and C must be controlled variables. The frequency of the note produced is the dependent (outcome) variable in the experiment.

492. (C) Comparing the frequencies for Group A (open at both ends) and Group B (closed at one end), one observes a significant drop in frequency.

493. (B) Pipe 2's frequency is over 20 Hz greater than all the other frequencies. Group A and B's data show that shorter pipes produce higher frequencies, so pipe 2 was probably too short.

494. (D) Table 12.7 shows pipe 3's frequency is 195 Hz. The only frequency close to this is when pipe 4 was closed at one end in Group B's experiment. The length of this pipe is 0.44 meters.

495. (D) Table 12.6 shows that Group A's open pipes had the higher frequencies. The data also show that shorter pipes have a higher frequency. The passage states that higher pitch corresponds with high frequency, so shorter pipes that are open on both ends will have the higher pitch.

496. (B) More or less water in the calorimeter will not affect the loss of thermal energy to the environment because the water is internal to the experiment. Both the draught shield and the insulating card will help prevent thermal energy from radiating into the environment. If the flame is close to the calorimeter, one would expect more of the thermal energy to transfer into the calorimeter as desired. If the flame was significantly separated from the calorimeter, one would expect a significant portion of the thermal energy to flow around the calorimeter, drafting into the environment.

497. (D) If twice the oil were completely burned, twice the thermal energy should be transferred to the calorimeter in an ideal experiment, and thus the temperature change would be greater. On the other hand, since the heat of combustion is the ratio between thermal energy transfer and mass combusted, the ratio would not change because thermal energy and mass are both doubled.

498. (A) When you subtract the initial temperature from the final temperature in Table 12.9, the greatest temperature change (5°C) is for the coconut oil.

499. (B) Since the same amount of oil was burned in each experiment and the same amount of water was in the calorimeter, the greatest temperature change would indicate the largest amount of energy transferred from the oil to the water. Ordering from largest to smallest temperature change in Table 12.9, one gets coconut, red palm, peanut/olive, and safflower. Ordering from largest to smallest percent saturation in Table 12.8, one gets coconut, red palm, peanut, olive, and safflower. Thus, the amount of energy transferred is greater for the oils that are more saturated.

500. (D) First, the amount of thermal energy transferred from the coconut oil to the water is calculated from $Q = m\,c\,\Delta T$ = (100 grams)(4.184 Joules/gram/°C)(26°C – 21°C). Next, to calculate the heat of combustion in Joules/gram, divide the thermal energy transferred by the mass of the oil burned:

$$\text{Heat of Combustion (J/g)} = (100\text{ grams})(4.184\text{ Joules/gram/°C})(5\text{°C}) \div (0.50\text{ grams})$$